M

The Physics of Hockey

The Physics of Hockey

ALAIN HACHÉ

The Johns Hopkins University Press
Baltimore and London

The Johns Hopkins University Press
2715 North Charles Street
Baltimore, Maryland 21218-4363
www.press.jhu.edu

Library of Congress Cataloging-in-Publication Data

Haché, Alain, 1970–
   The physics of hockey / Alain Haché.
      p. cm.
Includes bibliographical references and index.
   ISBN 0-8018-7071-2
   1. Physics.   2. Hockey.   3. Force and energy.   I. Title.
QC28 .H23 2002
530—dc21

2001008643

A catalog record for this book is available from the British Library.

*To all my teammates through the years
and to the championship 1996–97
U of T Physics hockey team*

Contents

Introduction

This book combines two very different elements—a science and a sport—each of which has its own attraction. The beauty of physics is its ability to reduce very complex phenomena to a few simple rules and equations. With these we can make predictions and understand how nature generally works, even though in reality there is great complexity. Because high school physics typically covers a small range of subjects, many people have a limited idea of what physics is used for. Some will tell you that physics is about finding out how long it takes for a falling stone to reach the ground or calculating the current through a light bulb. Although it can certainly be used to do that, physics is applicable to a much wider array of complex problems. To give an example, a team of scientists recently used physical models to predict the behavior of large crowds in a state of panic, like that which occurs when a fire breaks out in a building.[1] It is well known that the tremendous pressures caused by fans pushing against one another in an overcrowded stadium can have tragic results, such as in 1989 when 95 soccer fans were crushed in Hillsborough Stadium in Sheffield, England. Scientists made an important discovery with their physical model of crowd behavior: they found that strategically located columns near the exits could reduce the pressure and increase the flow of people through the doors. In other words, adding apparent obstructions could actually save lives. This counterintuitive finding was not derived from experience or trial and error (no one would dare set fires just to study crowd behavior), but rather it was the result of harmless simulations on a computer using models that obey simple physical laws. This is just

---

1. *Popular Science,* January 2001, p. 29.

one remarkable example of how physics is used to deal with seemingly unsolvable problems. After all, what is more complex and chaotic than a panic-stricken crowd? Likewise, much insight can be gained on the game of hockey using physical modeling.

The beauty of hockey seems quite different from what physics offers. Using a few simple game rules, talented athletes can turn the sport of hockey into an awesome spectacle full of unpredictable twists and turns. It is this unpredictability that makes the game so much fun to watch. From that viewpoint, physics and hockey appear to be at opposite ends of the spectrum, but, put together, they render each other service. Exploring that relationship is the main objective of this book.

Applying physics to hockey helps us understand how aspects of the sport work and lets us make use of that knowledge to improve our game. On the flip side, talking about hockey in a physics context may promote interest in science for the public at large and, by the same token, help create a better scientific culture (which many will agree is somewhat lacking in our society). I know this from experience, as students in my freshman mechanics class usually become interested when real-life examples are used, especially examples from hockey. Applications such as these make abstract theories come alive.

Physics has a long history of being successfully applied to sports. Over the last 30 years there has been an explosion of research in biomechanics and related fields. The knowledge has served, among other things, to develop better designs for equipment and help reduce sports injuries. The science and medicine of sports is now a well-established discipline with a strong academic presence in universities around the world. Already, physics books have been published on baseball, golf, skiing, running, sprinting, skating, and a whole slew of other activities. I hope that this book will help fill a void by discussing the physics of hockey.

Hockey may be fun to watch and study, but it's even more fun to play. During my humble amateur career, I've had a chance to play in many places and enjoy the company of people from all walks of life. One great thing about the game is that it makes us leave our cozy homes in the middle of the winter to meet and play with other people. I must also say I've had a lot of fun writing this book. As a physicist and a hockey player, I sometimes can't help but think about the game in a scientific way. Fortunately, hockey involves many facets of physics, perhaps more than any other sport. Because it is played on ice, we need to take into account elements of thermodynamics and molecular physics. Skating makes use of a great deal of mechanics, as does shooting. Puck trajectories are influenced by air drag and ice friction, which involve fluid dynamics. And because hockey is a contact sport, the physics of collisions is also part of the game.

In a way, my position as a goalie has given me a privileged view of the game. Half the time, when the action is away from my zone, I have the leisure to observe the game up close. This book greatly benefited from countless hours of observation done over the years. But of course, you don't need to be a goaltender or a scientist to enjoy a good game of hockey!

This book could easily have been turned into a boring scientific monograph. Instead, it is peppered with stories, real-life examples, and anecdotes, and should be accessible to a wide range of readers, including those with limited or no scientific background. I assumed the average reader would have nothing more than a bit of high school physics. However I did not want to leave out the scientifically inclined readers, so there is also enough mathematics to satisfy their appetite. Appendixes discuss the details of some important results covered in this book.

To many people, including myself, equations sometimes can be a turnoff. Though they are not the central part of this book, they are included to help the reader apply the knowledge gleaned to other situations. For example, someone may want to use the equation describing puck motion to estimate how much it has slowed down by the time it reaches the net after being shot from the red line. The shape of the equation is often as instructive as the equation itself: it tells us what parameters are important. In most cases, however, a reader can skim the formulas and go through the discussion without missing the main point.

An important point about physical units: because we are accustomed to speaking in terms of feet, miles per hour, and pounds, I often use this language for discussions in the book. Unfortunately, these units can create confusion if used in physics. So, unless stated otherwise, standard physics units such as meters, kilograms, and seconds are used in formulas. Refer to Appendix 1 for a conversion table of units.

The reader should keep in mind the limitations intrinsic to all physical models. Models are meant to be accurate to a certain extent, but, because simplifications are made, no model can reproduce a real-world situation with complete accuracy. For example, when the force of impact between two colliding players is calculated, the result is only an approximation of what happens in reality. Because hockey players are complicated three-dimensional bodies, there are many parameters that influence their impact. Each collision is different, and some cause injuries while others don't. In principle, there is nothing preventing us from creating more detailed models, but in doing so we would risk falling into tedious and boring details, which, in the end, would not give us much more insight. We can instead appreciate the essence of what goes on with simpler models.

The book is organized in a logical order of complexity. The physics of ice—the one main element of hockey—is the object of the first chapter. The unique combination of properties ice exhibits, especially its remarkable slipperiness, is an interesting topic in itself. Scientists have been mind-boggled by it for more than a century. In Chapter 2 we take a close look at the science of skating and the properties of the skate. There are different techniques commonly used by hockey players to move around on the ice. Also, the expenditure of energy needed to accelerate and to overcome air drag and ice friction is an important problem. Because one must shoot the puck to score a goal, Chapter 3 deals with the physics of shooting. We begin with the trajectory of the puck and the influence of the lift and drag forces caused by air. A model for the slap shot is provided, which helps us understand the important elements in achieving the highest speeds. The physics of collisions and injuries, a gripping subject, is explored in Chapter 4. There we analyze body checks against the board and at mid-ice using the laws of physics and obtain estimates of the force of impact. In Chapter 5 we discuss the science and principles behind the art of keeping the net. There are many skills involved in stopping pucks: good reaction times, proper positioning, largest cross section, and anticipation. In Chapter 6, we look at the game as a whole and the odds of winning, drawing on the concepts of statistics and probabilities.

Some of the material in this book is sequential. In other words, the text makes reference to previously discussed concepts. So readers who don't start at the beginning may occasionally need to refer to earlier chapters.

I wish to thank all the people who contributed in making this book a reality. I am especially grateful to Trevor Lipscombe, for initiating this project and for his many suggestions, and to Celestia Ward for her careful copyediting. I am also indebted to Oskar Vafek, Venitia Joseph, Daniel Côté, and Thomas Richard, who have proofread the manuscript.

The Physics of Hockey

Chapter 1

ON ICE

Hockey would not have the same appeal if it were played on ground or grass. Ice is what gives the "coolest game on Earth," as the NHL calls it, its distinguishing characteristics. For one thing, on any surface other than ice, it wouldn't be as fast. Although variations like roller hockey have appeared recently, they haven't reached the same level of popularity as ice hockey. Field hockey has a wide following, but mostly among Commonwealth nations.

Games on ice have existed for centuries. When the snow and cold weather enveloped towns and villages across the northern hemisphere, life didn't stop altogether. People created new amusements and sports, and ice became a favored playing field. Throughout history, humankind has found something thrilling about running, skating, and playing sports on a slippery surface. There are references to ice games in Europe from as early as the Middle Ages. Soon after the arrival of the modern skate in the nineteenth century, people found that the next best thing to skating was shooting a ball or a puck across the ice with a stick. Thus hockey was born, and, since then, generations of children have grown up playing this favorite sport on frozen lakes and ponds. After countless transformations and improvements, it has become the world's most popular winter team sport.

So exactly when and where was hockey invented? There are many claims but no consensus. Because many regions in the world have natural ice in the winter, it is not surprising that we hear different stories about where hockey began. At least one thing is certain: Canada is where hockey first evolved most fully and gained the most popularity, making that country the main supplier of NHL talent until today. Hockey is now such a part of the Canadian psyche that when Marc Garneau became the first Canadian astronaut to fly in space, in 1984, he brought with him a hockey puck, of all things.

Quebeckers claim hockey was first played at McGill University in Montreal, whereas Ontarians would rather believe it started in Kingston, Ontario. The city of Halifax in Nova Scotia also lays claim to the honor of invention. About the only undisputed fact is that the name *hockey* came from the French word *hoquet,* meaning "shepherd's crook." However, many historians also believe that the earliest form of the game appeared in the early nineteenth century, when eastern Micmac Indians played a sport that combined elements of Native American lacrosse with the Irish sport of hurling. The game was played with hurley sticks and square wooden blocks.

Hockey as an organized team sport spread through Canada in the mid-1850s, and the first league was created by four clubs in Kingston. It didn't take long before rules were codified and tournaments were organized everywhere. In 1893, the now famous Stanley Cup—named after Frederick Arthur Stanley, Lord of Preston, sixteenth Earl of Derby, and Governor General of Canada—was introduced as an annual award for hockey excellence in Canada. The National Hockey League was founded in Montreal in 1917, and, starting that year, the trophy was awarded to the NHL playoff champions. The first championship team was the Toronto Arenas. The arrival of artificial ice and indoor rinks in the early twentieth century made longer hockey seasons possible and helped spread the sport to warmer regions of the world. As many Canadians will tell you, this increased popularity was a double-edged sword. Canada had been home to the NHL's best teams, but over the years teams such as the Quebec Nordiques and the Winnipeg Jets moved south of the border to more lucrative markets. This was hard to take for a nation that, according to a recent poll,[1] considers its series win against the Soviet Union in 1972 more important than its participation in the Second World War! The Minnesota North Stars have also joined this southern migration, moving to Dallas, although another Minnesota team (the Wilds) has been recently resurrected. Yet artificial ice

---

1. *Toronto Globe and Mail,* September 18, 2000, p. 1.

technologies have made possible teams like the Phoenix Coyotes and the Dallas Stars, whose host cities have average daytime temperatures of 66 and 55°F (19 and 13°C) in the middle of January. There are also benefits in terms of international competitions. Countries where ice is only found inside freezers now field teams at tournaments. Nations like Italy, Korea, and South Africa now have leagues of their own, and it may only be a matter of time before a warmer country wins an Olympic medal.

From a physics point of view, ice is a fascinating subject because of its unique and somewhat bizarre properties. Surprisingly, ice is very much a contemporary topic of research among scientific circles. For example, the mechanism responsible for one of ice's most fundamental features, its slipperiness, was only recently unraveled. The physics of ice is an active field of research that is important not only to hockey and winter sports but also to chemistry, engineering, geology, and oceanography. In our quest to understand ice, we will start with the simplest question of all. If you want to play hockey you need ice, so how do you make it?

The Ice Cometh

Water, as everyone knows, freezes at 0°C. So a simple plan to make an ice rink would be to fill a space with water and cool it down. It may sound trivial but in practice it is not, owing to the huge amount of energy involved.

To understand why, let's first take a look at the physics of cooling. The coldest anything can get is a temperature known as *absolute zero,* which is a tad below −273°C. While the January chill may make it painfully cold to wait in line outside for a ticket to an Oilers' game in Edmonton, it's rare for outdoor temperatures to get below −50°C. According to *The Guinness Book of World Records,* the coldest outdoor temperature ever recorded was −89°C, in Vostok, Antartica (which is still quite balmy compared to absolute zero). Anything above −273°C contains some heat, or thermal energy. The amount of thermal energy contained depends on the temperature, the mass, and the stuff of which the object is made. A gallon of water at 20°C has more thermal energy than a gallon of water at 10°C. And two gallons of water at 15°C have twice as much thermal energy as one gallon of water at the same temperature. When scientists and engineers developed the steam engine in the nineteenth century, they became fascinated with turning liquids into gases and solids. Developing an understanding of these phase changes, as physicists now call them, was a necessary step in making better engines. A crucial problem was figuring out exactly how much energy it took to raise or lower the temperature of a given substance, say water, by one degree Celsius. This quantity is called the substance's *heat capacity;* for water this amount is around 4.2 joules/gram/°C. In other words, for every gram of water that we want to cool down by one degree, we have to take away 4.2 joules of energy. We can now imagine how much energy is needed to resurface the ice at Madison Square Garden for a Rangers game using warm water. Making ice from scratch is extremely costly. The size of a hockey rink is about 1,600 square meters, so filling it to a depth of 2 cm requires 32 million grams of water. If the water is initially at room temperature, the thermal energy that must be extracted to cool this much water down to 0°C is some 2.7 billion joules, enough energy to power an average house for two weeks. It would take weeks of steady work by a typical household refrigerator just to cool the rink at Madison Square Garden down to the freezing point. And that's assuming the water is insulated from its environment, which isn't the case. Fortunately, as we will see later, there are systems better than regular appliances to cool hockey rinks.

Water has a rather large heat capacity compared to most liquids, and that's why it is a popular choice for cooling (or heating) systems, whether in thermonuclear reactors, car engines, or air conditioners. Water also has a huge influence on climate. Large bodies of water like oceans and lakes have a stabilizing effect on coastal climate because they act like massive heat reservoirs, which explains why Chicago has more drastic temperature fluctuations over the year than New York. Unfortunately for those maintaining ice rinks, water's large heat capacity means that creating ice involves a lot of energy.

So assume we have used 2.7 billion joules of energy to cool down our rink to 0°C. But we don't have ice yet. Why? Because it takes additional energy just to change a liquid into a solid, even if the temperature stays the same. This amount of energy is called the *latent heat of fusion,* and for water it is 340 J/g. The mass of water we've flooded the Garden with now needs an extra 11 billion joules just to freeze! For our poor household fridge, this would mean a lot of extra hard work.

Temperature is not the only parameter that changes during cooling: density also varies, and in a very peculiar way for water. Water has a density near 1 g/cm^3 (which was at one point used to define the gram), but, like most liquids, its density tends to increase at lower temperatures. However, at 4°C the process is reversed and water's density drops gradually to 0.99984 g/cm^3 at 0°C. When freezing occurs, density drops further to 0.9167 g/cm^3, which allows ice to float with 8 percent of its volume rising above the liquid water surface. Saltwater is heavier than pure water, so icebergs will float with about one-tenth their volume above the ocean—the proverbial tip of the iceberg. Water is one of the few substances that expand upon freezing; other liquids do the opposite. If water followed the same rule, icebergs would sink and the Titanic disaster would never have happened.

This peculiar density behavior of water plays a role in the formation of ice on lakes and ponds, the surface on which so many grow up playing hockey. Going from your basement to the attic on a warm summer day, you'll notice the temperature will increase, because hot air rises. The same is true for a lake: as temperatures drop during the fall and winter, the cooler and denser water remains at the bottom and the warmer water rises to the top. But once temperatures fall below 4°C , the situation changes. The colder water rises to the top and stays there. This inversion process is called the "turning over" of a lake. Because of this, natural ice only forms once all the water in the lake or pond has reached 4°C or colder. Eventually the top layer becomes cold enough and turns to ice. This means lakes and ponds freeze from the top down, creating a layer that grows downward. This phenomenon is behind the warning that you may be "skating on thin ice"—and why anyone venturing onto a seeming frozen lake must be very careful.

The Water Molecule: The Lego Block of Ice

Water has an interesting combination of properties: it is a good solvent, it has a large thermal capacity, it is chemically stable (that is, it doesn't light up easily!), and it is in good supply. It also happens to freeze within the range of temperatures at which life is possible. If water did not have these properties, there would be no life—and no hockey. So far in our look at how ice is formed, we've used a nineteenth-century approach: heat is simply extracted until water freezes. But if we really want to find how to make perfect ice (or at least the best ice possible), it helps to take a look at the molecules that compose ice. Richard Feynman, one of the greatest and best-known physicists of the twentieth century, once made this important point: if all scientific knowledge were to be lost except for one principle, he'd like that principle to be the atomic hypothesis, which says that all matter is made of atoms. It's that important of a concept.

If there's one chemical formula most people remember from high school, it is H_2O. A water molecule (see Fig 1.1) is made of two hydrogen atoms and one oxygen atom. Hydrogen is the most basic atom around, with one positive proton at its core and one negative electron circling on the outside. Oxygen is far more complex, with eight protons in the nucleus and eight electrons arranged in shells around it. The water molecule is roughly shaped like a boomerang: a central oxygen atom is flanked by smaller hydrogen atoms located on either side, forming a 105° angle. The diameter of the molecule is only about 0.5 nanometers, or less than a millionth of a millimeter. The force holding the water molecule together is described as a covalent bond, in which the three atoms share electrons and become closely linked. This tendency to share electrons stems from the fact that atoms are "happiest" (that is, most stable) when they have a complete set of electrons in their outer shell. The reason behind this has to do with quantum mechanics—it can't be explained using classical physics. Those atoms with completely filled outer shells don't

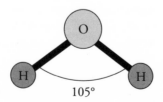

Figure 1.1. The water molecule consists of two hydrogen atoms and an oxygen atom joined by covalent bonds. Its size is a mere 0.5 nanometers.

really need to combine chemically with anything else; these are the so-called inert elements, such as argon, xenon, and krypton. Oxygen and hydrogen, meanwhile, need two and one extra electron respectively, hence the happy threesome occurring in each water molecule.

There is no overall electric charge to a water molecule, as all atoms inside are neutral. But electrons, quantum-mechanically speaking, are not evenly distributed throughout the molecule, so some regions end up slightly positive and others slightly negative. In other words, the water molecule is *polarized.* This important aspect is responsible for some of water's unique properties. Polarization affects how neighboring molecules interact. As the positive part of one water molecule is attracted to the negative part of another, they stick together. In scientific jargon this is called *hydrogen bonding.* While they are fairly weaker than covalent bonds inside the molecule, hydrogen bonds are behind some of water's characteristic properties, such as its large heat capacity.

When two water molecules are attracted to each other, it takes some energy to keep them apart. One source that can do this is thermal energy. Above 100°C, there's enough thermal energy to break the hydrogen bonds completely, and each molecule goes its own way. Below 100°C, water molecules take up a more limited space, rubbing against neighbors while not being attached to any of them—that is, they are in a liquid state. Liquid water is therefore much denser than vapor, but it still retains fluid properties. As water is cooled below 15°C, the lower thermal energy allows molecules to stick even more closely together, promoting the formation of clusters and chains called *polymers* (see Fig. 1.2). At 10°C the long wiggly chains contain ten or so molecules. This clumping explains why the density starts to decrease below 4°C. As more molecules join these individual parties, space tends to form between clusters.

Polymers are an example of what physicists call *short-range ordering.* Beyond a distance of a few molecular diameters, the positions of molecules are not related anymore, just like magnets cannot interact when moved apart. At 0°C the vibrational energy in water is small enough to make long-range ordering possible, so that all molecules can act in concert and crystallize. But making the jump from a disordered (liquid) to an ordered (solid) state comes at a price. For the molecules to arrange themselves into a perfect array, some molecules

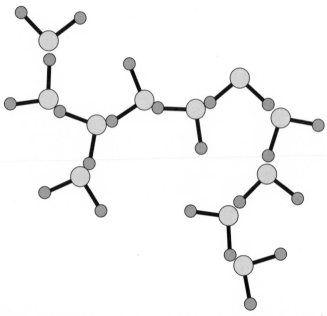

Figure 1.2. Polymerized water consists of a chain of water molecules. Clusters are held together by hydrogen bonds and may contain a dozen molecules or so depending on the temperature.

must fit into new spaces and new bonds must be formed. The additional energy released from the formation of these bonds must be removed in order for ice to form, which explains the latent heat of fusion discussed earlier. Even when the temperature falls far below freezing, ice will not form as long as this extra heat remains. Indeed, liquid water can exist below 0°C, though it has to be fairly pure and conditions need to be just right. This is an example of what is called a *supercooled,* metastable liquid. By the same token, water can exist above 100°C without necessarily turning into steam. When superheated water does turn into steam, it does so with a bang, like in a kernel of corn when it explodes into popcorn. Metastable liquids occur in nature as well. If tree sap could not exist in liquid form well below its official freezing point, no forests would exist in Canada and other places in the "frozen North."

When ice forms, the molecules arrange themselves a bit differently. One question physicists like to ask is, How many neighbors will a single water molecule have, or how "sociable" is it? In all, a water molecule can take no more than four companions: its oxygen atom can bond with two more hydrogen atoms (one on each side), while its own two hydrogen atoms can link with the oxygen of two more molecules. These five molecules form a tetrahedron, a shape popular in elementary geometry textbooks. This configuration is the basic unit of ice's crystal structure (Fig. 1.3), which looks like a web of interlocking hexagons. Incidentally, this hexagonal structure is responsible for the starlike shape of snowflakes, as the first few molecules to crystallize align themselves along a tetrahedron and the rest follow.

An ice crystal doesn't look the same in all directions, however. Layers of molecules compose sheetlike structures. Molecules within a sheet are more tightly bound together than are those in two adjacent layers. It is therefore not surprising that the distance between layers is greater than that between hexagons: 0.734 nm and 0.452 nm, respectively. As we will see later, this sheetlike configuration is an important element in understanding the slipperiness of ice.

Hexagonal ice is not the only possible type of crystallization. More exotic forms, such as cubic ice, have been observed, although only under combinations of pressure and temperature that would not be present on the hockey rink.

The Phase Diagram

Learning about attractive forces between water molecules helps us understand what happens when we heat and cool water. Another condition that plays a critical role in determining whether water remains in liquid form or not is pressure. At normal atmospheric pressure (101.3 kilopascals, or kPa, otherwise known as 1 atmosphere, or atm), ice and vapor form at 0 and 100°C, respectively. But these boiling and freezing points change with pressure. For example, when New Zealander Sir Edmund Hilary and Sherpa Tenzing Norgay of Nepal climbed Mount Everest in 1952, the lower atmospheric pressure at 29,000 feet caused water to boil at less than 70°C, prompting Hilary to complain that they could not make a decent cup of tea. Likewise, pressure cookers are popular in Denver, the "Mile High" city, because residents can more easily get well-cooked meals by increasing the pressure and therefore raising the temperature at which

(a)

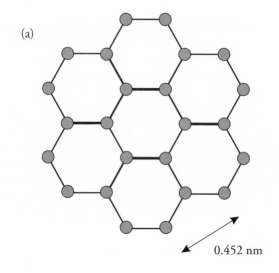

0.452 nm

(b)

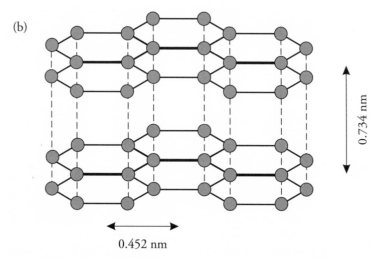

0.734 nm

0.452 nm

Figure 1.3. The crystalline structure of ice is hexagonal, like graphite, with sheets of closely packed molecules superimposed on top of one another. Bonding is stronger within a sheet than between them. (a) and (b) show the top and side views of ice's structure, respectively.

water boils. When it comes to hockey, the pressure existing between the ice and the skate blade is considerable and should not be overlooked, as it may liquefy the ice by changing the melting point.

Since both temperature and pressure determine whether water is in its solid, liquid, or gaseous phase, comprehending how these environmental conditions act altogether can be difficult to grasp. This is why physicists developed the *phase diagram*. In a controlled experiment, we can fix the pressure on a sample of water, then measure the temperature at its melting and boiling points. We can then change the pressure and repeat the process. By "joining the dots" and drawing two curves, one representing the changes from water to ice and another representing changes of water to steam, we come up with the phase diagram shown in Fig. 1.4. With such a chart in hand, given a specific temperature and pressure, we can determine whether the water under those conditions would be solid, liquid, or vapor. For hockey fans, the interesting part of the curve is around 1 atmosphere of pressure and near 0°C. In this region, when you increase

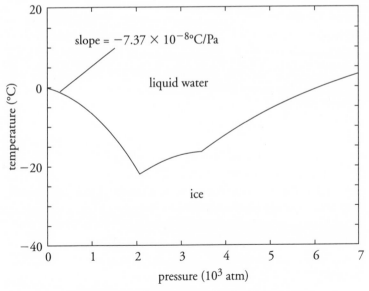

Figure 1.4. Phase diagram of pure water at the liquid / ice boundary. At normal pressures, the freezing point decreases with added pressure.

the pressure the melting point drops, which means you can melt ice below the normal freezing point simply by applying pressure. More precisely, each pascal of added pressure (the pascal, which equals 1 N/m^2, is the standard unit of pressure, named in honor of French physicist Blaise Pascal) reduces the melting point by 7.37×10^{-8}°C. That is not a whole lot, but one pascal is not much pressure either, only 0.00015 lb./in². A 250-pound hockey player like Ottawa defenseman Zdeno Chara, standing on one skate, can put much more pressure than that on the ice under his blade, thereby significantly reducing the melting point. We'll see later how this pressure-induced melting may help the skating process.

Having covered the physics and chemistry of water and understanding how it is converted into ice, we now turn to something more applied: the technology that makes it possible to create and maintain high-quality, NHL-caliber ice rinks.

The Art of Making Ice

Creating an ice surface is one thing, but to make and maintain a top-quality indoor rink is quite another. Today's ice-making technology and techniques are the product of decades of trial and error. Around the world, the art of making and marking ice for hockey purposes has become a trade employing thousands of arena workers. Despite what many may think, it's a job requiring a great amount of skill and knowledge. Keeping things under control, no matter what the outside temperature might be, demands nonstop care from ice-keepers and machines. As one might suspect, modern arenas are equipped with advanced refrigeration systems to keep the ice slab chilled. Huge compressors refrigerate and pump the cooling fluid under the rink. But even in the twenty-first century, the cooling fluid is far from high-tech—it's saltwater. Brine, as it's also called, freezes well below 0°C and, since it is still mostly water, has a large heat capacity that makes it quite efficient at freezing the ice.

The game plan is simple: to cool the rink down you transfer its heat to the brine. The saltwater is cooled to subzero temperatures and carried through an extended network of crisscrossing pipes underneath the ice surface, inside a thick slab of concrete. Upon contact, the brine heats up and the slab cools down. This is what engineers call a heat exchanger, and it is the same principle used in cooling your car engine. Coolant circulates around the engine block, taking heat away from the engine and sending it to the radiator and then to the outside world.

Once the concrete slab is cold enough, the first few layers of water are spread on it and ice-making begins. The first layers of ice are sealed with a layer of white paint to increase the color contrast between the puck, the lines, and the ice. Without the paint, the ice would look grayish, which it did in the old days. More layers of ice are then added, and, once it has reached a thickness of about one-eighth of an inch, lines, dots, circles, logos, and advertisements are applied using water-based paint that dries quickly. Nowadays, a special kind of durable, paperlike tissue is often used to make straight lines. The paint is then covered with another eight to ten thin layers of ice. This process ensures that the marks are well inside the ice and do not disappear after the first resurfacing. With time, though, skates dig deep enough to remove and damage lines bit by bit until they become dim and need repainting. Making a standard-size rink from scratch is a matter of one or two full days of work.

So many layers of ice may sound like a lot, and, indeed, as much as 40,000 liters goes in to icing a standard rink. But, unlike the ice on lakes and rivers, the final thickness of the ice is less than one inch; typically it is only three-quarters of an inch to one inch. While you wouldn't want to play on such thin ice on a lake without wearing a life jacket, indoor ice is entirely supported by the concrete slab, so there is no reason for it to be thicker.

Depending on the sport being played, the ice temperature may be adjusted from about −5°C to −10°C. The icemakers at the San Jose Arena, home of the San Jose Sharks, keep the ice at −5.5°C for figure skating and −9°C for hockey. Other rink managers prefer slightly warmer ice, but cooler ice is harder, which makes it better for fast skating. Figure skaters prefer a softer ice surface for smoother landings, whereas hockey players like it hard and fast. Maintaining the appropriate ice temperature throughout a game in a place like the America West Arena in Phoenix, with 16,000 cheering fans, is a challenge. Humans burn, on average (accounting for different body sizes and children), about 1,000 calories a day. A good part of this energy is transformed into heat and released into the environment through breathing and through the skin. During a two-hour hockey game, we can expect the Coyotes fans to burn roughly 1.4 million calories, or nearly 6 billion joules (probably more if the game is tight and they're getting excited). That's plenty of energy to heat up the place.[2] The humidity coming from the outside environment adds to the problem. Since humidity is water, it is very effective at warming things up, and a rise of a single degree in temperature will make the ice more susceptible to chipping and damage.

With so many people packed in an arena, keeping the ice cool becomes a real challenge, especially in southern regions where heat is a year-long problem. Every time deliveries of beer, pretzels, or hot dogs arrive for the concession stands, the doors are opened and heat and humidity come in. Though this is hardly a concern in places like Ottawa and Edmonton in the middle of winter, many arenas in the South have dehumidifiers working at full capacity during the game. All things considered, it is quite amazing that the Florida Panthers, in Miami, can play at home in June! Without modern ice-making technology, the NHL would never have been able to expand south and the regular season would not be 82 games long, even in Canada.

In all NHL arenas the ice is kept at least from September until May, and sometimes all year round, even though other events take place. For example, the newly built Air Canada Center in Toronto is home to the Maple Leafs as well as the Toronto Raptors, an NBA franchise. When the Raptors are in town, a couple of hours is all the crew needs to install a basketball court on top of the ice and arrange the seats around it.

A Cool Guy Called Zamboni

Over the course of a 20-minute period in a hard-fought hockey game, the ice suffers a considerable amount of damage. As skates leave deep traces and the surface is scraped, tiny bits of ice quickly accumulate, making the puck bounce and slow down. Goalies constantly sweep the front of their nets to get rid of excess snow and thereby reduce

---

2. This concept of "animal heating" has even been exploited to heat houses—using the energy produced by rabbits, most of which is released through their ears.

the risk of unpredictable puck trajectories. When the buzzer rings, it's time for players—and the ice—to take a much-needed break and let the marvelous machine officially known as the *ice-resurfacer* but almost always called the *Zamboni* smooth out the ice.[3] Doors open, nets are pushed aside, and the boxy vehicle makes its entrance, leaving a trail of vapor in its wake. Less than ten minutes and a few laps later, the ice is shiny, smooth, and ready to take another period of beating. At some rinks, two Zambonis do the sweeping, and it's not uncommon for "Zamboni races" to take place, to see which driver can resurface their half of the rink first.

The name *Zamboni,* synonymous with ice-resurfacer, comes from its inventor, Frank J. Zamboni (1901–88), a tireless inventor and entrepreneur from southern California (of all places), with expertise in refrigeration. As owner of his own indoor rink, Frank Zamboni was faced with the challenge of maintaining a good skating surface in a region of the country where the climate is hostile to ice. In the 1940s, resurfacing the ice meant pulling a scraper behind a tractor to shave the surface. It took up to four workers to scoop away the shavings, spray water over the surface, and squeegee it clean; including time to allow the water to freeze, the process took over an hour. The technological challenge of automating the whole operation became Zamboni's obsession, but it was destined to be an adventure with many trials, setbacks, and countless improvements. Eventually, Frank Zamboni's efforts proved fruitful. In 1949 the Model A Zamboni was born, and, for the first time, not only was a single machine able to consistently produce a good sheet of ice, but it could do it in much less time.

Mr. Zamboni's first ice-resurfacers did not look like the modern ones: they were marvels of craftsmanship, built using bits and parts of different vehicles. The first Zamboni to closely resemble the ones we know today was the Model C, made in 1952. Built on a complete Jeep, the vehicle had an elevated driver seat at the rear and a lower snow tank (a reservoir in which the ice chips are dumped) to allow better visibility. The basic design of the Zamboni has remained the same ever since, although many technological improvements have

---

3. For further information on ice resurfacing, the Zamboni, and its history, consult the company's website at www.zamboni.com.

been made. For example, some machines today are entirely powered by an electrical motor, eliminating the carbon monoxide that is produced from gas-powered engines.

The role of the ice-resurfacer entails more than just sweeping up the snow and pouring warm water on the ice. Producing a good sheet of ice is a complicated process that involves four steps, though modern Zambonis do these steps all in one sweep. Fig. 1.5 shows the inside of a modern Zamboni. Located underneath the driver's seat is a long, sharp steel blade that removes a thin layer of ice. This evens out the surface and helps remove the grooves. In the second step, an auger removes the shavings and sweeps them up into a vertical screw, which in turn dumps them into a large bucket for storage. The bucket is later emptied once the job is done. The next stage involves washing the ice, in order to flush the grooves and loosen any dirt or debris that has accumulated on the surface. For this purpose, water from a smaller reservoir is sprayed evenly on the ice, and the excess water is squeegeed and vacuumed out. To avoid unnecessary waste and to keep the size of the reservoir to a minimum, the wash water is filtered and recirculated. Finally, a coat of water is applied uniformly with a pad, which quickly freezes and creates a smooth, shiny surface. The warm water, usually heated to around 60 to 65°C, "fuses" with the ice quicker than cold water would. This helps melt the remaining irregularities and replaces the original layer that was shaved. The warmer the water, the more even the new surface will be.

Although driving a resurfacer around seems easy and fun, it's not that straightforward. Operators are usually required to have a driver's license and take some training lessons. Even so, rink owners will usually only trust someone enough to hand them the keys to this vehicle after a couple of years of supervised experience. It's too easy to destroy a rink with that machine to let a novice drive it. For example, because of the Zamboni's sheer weight, stopping the machine for too long could create holes in the ice. There are many tricks to be learned, such as determining where the ice is thinner based on how faded the lines are. A skilled resurfacer needs to know how to add water to thicken those areas.

Today's Zambonis are used on ice rinks around the world and smooth out ice surfaces for prestigious events such as the Winter

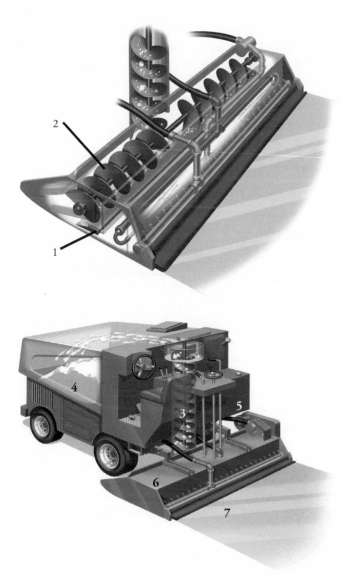

Figure 1.5. Inside view of a typical Zamboni ice-resurfacer. A sharp blade shaves the ice (1), then shavings are collected with a horizontal screw (2) and propelled upward with a vertical screw (3) into the snow tank (4). Water from another tank (5) is poured into the conditioner (6) to wash the ice. The dirty water is vacuumed out and a final coat of warm water is applied to the ice with a pad (7). Courtesy of Frank J. Zamboni & Co.

Olympics and NHL games. They are sophisticated machines weighing up to 4 metric tons when full of water and have a capacity of up to 1,000 liters, enough to fill 2 bathtubs to the brim. Thanks to the modern ice-resurfacers, hockey fans can watch their favorite sport without having to wait hours between periods.

A Word on Friction

Most people's first experience on ice, whether it's at a frozen lake or an ice-skating arena, consists of falling over. The conclusions of this experiment, if we may call it that, are: (1) ice is hard, and (2) it is slippery. Although we can certainly shatter our bones on it, ice is not among the hardest materials. On a scale of 0 to 10 (10 being diamond, the hardest thing around), ice has a hardness of less than 2, whereas most solids are in the 3–7 range (glass is at 7). Of course, when hockey players talk about "hard" and "soft" ice, they are referring to relatively modest changes in the ice properties, which nonetheless affect the performance of the skater. As we will see later, ice's hardness influences its slipperiness.

Slipperiness is related to a property physicists call *friction*. Although friction has been studied for a long time, it is still not clear how it works exactly, at least when analyzed at an atomic scale. Luckily, for the purpose of this book, we don't need to worry about the fine print. We know that friction occurs in nature whenever two bodies are in contact. Nothing is perfectly smooth, so there are always small indentations and irregularities on their surfaces, even if this roughness is microscopic. When two solids rub against each other, the microscopic irregularities grind against each other and provide resistance to motion—friction. This force has some important characteristics worth mentioning at the onset.

1. Friction is always antagonistic (a trait some hockey players can relate to). That is, it always opposes motion, never helps it. It is oriented in the exact opposite direction of motion (or velocity) and therefore slows things down.
2. It is parallel to the plane of contact. The nature of roughness only creates a force that is parallel to the surface, so it can't push things apart or pull them together.

3. It can be *static* or *dynamic:* friction can exist whether the body is at rest or in motion. At rest, friction counteracts forces that tend to move the body. For example, when you give a small horizontal pull on a sled at rest, if the friction force is equal and opposite in direction to the pulling force, the sled won't budge (the net force is null). Therefore, static friction may be as small as zero or as large as the force tending to cause the motion. If you pull hard enough and move the sled, the friction loses out but remains constant and opposed to the movement; this is the dynamic case.
4. Friction depends on the force of contact, or the pressure between the two bodies. Intuitively it makes sense that the harder we press two objects together, the harder it will be to rub them.

Experimentally, it has been determined that the friction force is directly proportional to the force of contact, labeled N—also called the "normal force" because it is perpendicular to the plane of contact. Mathematically we can write:

$$f = \mu N, \tag{1.1}$$

where f is the friction force and μ is a constant called the *friction coefficient.* Usually the dynamic coefficient of friction is slightly lower than the static one. This is why it is harder, for example, to get the hockey net moving from rest than to keep it going.

The friction coefficient depends on the two materials that are in contact. As one might expect, rubber on ice has a smaller coefficient than rubber on asphalt, and skating with rusty blades is not as easy as with clean skates. To give an idea of the typical range, the list on the next page shows the friction coefficient for a few materials.[4]

A common example of friction involves a body moving on a horizontal surface. In this simple case, the friction force depends solely on the weight of the object. Let's take a look at Fig. 1.6, which shows the various forces acting on a puck in motion on the ice. First, gravity produces a downward force, the weight, represented as *mg*, with *m*

---

4. Taken from D. C. Giancoli, *Physics,* 4th ed. (Englewood Cliffs, N.J.: Prentice Hall, 1995), 93.

Materials	dynamic friction coefficient
rubber on concrete	0.8
steel on steel	0.6
wood on wood	0.2
wax on snow (i.e., skiing)	0.1
steel on ice (i.e., skating)	0.005
ice on ice	0.003
joints (knee, elbow, etc.)	0.003

being the mass of the puck in kilograms and g a constant called the *acceleration due to gravity*, which is about 9.8 m/s^2 at sea level. (Although people commonly use the terms *weight* and *mass* interchangeably, in physics they are not the same. Weight is the force of gravity on a body, measured in newtons.) Because the puck doesn't move up and down, the net vertical force on it must be zero, so the ice is "pushing upward" with the same force that gravity is pulling it down with. This is the *normal* force between the ice and the puck, and we can write $N = mg$. The friction force works against the velocity of the puck and has a magnitude of $f = \mu N = \mu mg$. The coefficient μ is determined experimentally by taking the ratio between the friction force and the weight of the object moving horizontally. Since gravity and

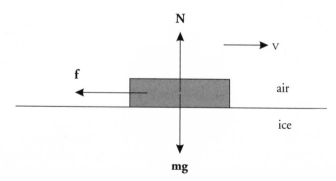

Figure 1.6. The forces acting on a puck moving on the ice surface. The normal force (N, the reaction of the ice to the weight of the puck) counteracts the weight mg exactly. The net resulting force is simply friction f causing the puck to slow its velocity v.

the normal force cancel out, the net force is the friction force, which will cause the puck to slow down. The equation of motion (see Appendix 2 for a review of basic mechanics) is obtained from Newton's famous second law, $F = ma$, and is therefore $F = \mu mg = ma$, or simply $a = \mu g$. This is the acceleration, or, more properly, the deceleration of the puck, which is typically less than 1 m/s². I should point out that if the puck is moving on an inclined plane instead of perfectly horizontally, these equations would no longer apply and the angle of inclination would need to be taken into account.

Slippery When Wet

In terms of friction, ice is in a class of its own, as the list of friction coefficients above suggests. Friction is much smaller on ice than on common materials such as plastic or wood. This is the single most important property of ice as far as hockey is concerned, since without its low friction coefficient skating would be impossible. But why is ice slippery in the first place? Amazingly enough, the clear scientific answer to this simple question has only been provided recently. For a long time scientists have known that the slipperiness of ice was caused by the lubricating action of a thin film of water between the two contacting surfaces. What produced this layer of near-frozen water was, however, hotly contested. Traditionally, there have been three explanations put forth to explain the melting: pressure, friction, and another mechanism that does away with the concept of melting altogether. Let's look at them individually.

1. Pressure melting

As mentioned before, the temperature at which ice melts drops when it is under pressure. Technically speaking, and as the phase diagram of water in Fig. 1.4 shows, at one atmosphere of pressure the phase boundary between liquid water and ice has a negative slope. In other words, any extra pressure reduces the melting point.[5] If pressure is subsequently reduced, water will freeze again. This process is called *regelation* and was discovered by English physicist Michael Faraday.

---

5. S. C. Colbeck, "Pressure Melting and Ice Skating," *American Journal of Physics* 63 (1995): 888.

A neat experiment based on regelation involves hanging two heavy masses connected by a thin wire around a long block of ice supported at both ends. Because of pressure melting, the ice under the wire liquefies and the wire cuts into the block. As the wire sinks through, the water is pushed above it and, no longer under pressure, freezes instantly. The result is a wire that seems to magically penetrate the solid ice. Eventually the wire cuts completely through and the masses fall away, leaving the ice block intact. Even Houdini would have been impressed!

The idea behind explaining slipperiness by means of pressure melting is that a heavy player on a pair of narrow skates exerts an awful lot of pressure on the ice. This would cause the surface layer to melt so the player glides on a thin film of water. Once the skater has moved, the pressure is gone and the water left behind the blade turns back into ice again. But is this believable?

The problem is, in order to cause an appreciable drop in the melting temperature, pressure has to be extremely high. According to our phase diagram, reducing the freezing point by one degree centigrade requires a pressure of 14,000 kPa, or 130 atmospheres! Even Eric Lindros, at 230 pounds, doesn't come close (NHL hockey players are getting bigger, but they are not to that point just yet).

The part of the blade that touches the ice is roughly 3 mm wide and 10 cm long, giving a total area of 6 cm^2 for two skates. A 90-kg hockey player standing on two skates would therefore create only 1,460 kPa of pressure. Although you wouldn't want to get caught under the blade of someone exerting this much pressure, it is hardly enough to make a dent on the melting point of ice—it lowers the melting point by about 0.1°C. It therefore seems safe to say that pressure melting is only a factor when the ice temperature is very close to 0°C. Moreover, taking a close look at the part of the phase diagram below −20°C, we see that no amount of pressure can produce pressure melting. At that temperature, the slope on the phase boundary is positive: adding pressure increases the melting point. Because ice is slippery even when little pressure is applied (think of a puck sliding) and at very low temperatures (I've skated outside at −25°C without problems), we can rule out pressure melting as the main cause for slipperiness.

2. Frictional Heating and Melting

Just as rubbing your hands together will produce heat, moving a skate across an ice surface will raise the temperature between blade and ice. The faster the movement and the larger the pressure against the ice, the more heat is generated. Hence we have the idea that slipperiness is caused by a layer of water formed when the ice at the top absorbs this excess heat.

How much thermal energy is really created when you rub a skate along the ice? To answer that question, we start by calculating how much energy is released at a particular point on the ice. The friction force between the blade and the ice is $f = \mu mg$, assuming the weight is evenly distributed over each skate and the player is simply gliding on the ice. The amount of energy (heat) this force produces depends on the distance over which the rubbing takes place. When the blade, which contacts the ice over a length L, completely sweeps over a given point on the ice, it leaves behind the following amount of heat:

$$E = fL = \mu mgL. \tag{1.2}$$

Now suppose that the ice absorbs roughly half the heat, while the rest is dissipated through the blade and elsewhere. As a result, the ice temperature will rise by an amount ΔT. By how much will it increase? According to the definition of heat capacity (noted as C), the temperature change also depends on the mass, given by ρV, where ρ is the density and V the volume of the ice absorbing the heat. Therefore,

$$\Delta T \approx \frac{1}{2} \frac{\mu mgL}{\rho VC}. \tag{1.3}$$

Somehow we need to work out what the volume V is, however. If the area under the blade is A, then $V = Ad$, where d is some penetration depth to which the heat has diffused. To find d, we need another piece of nineteenth-century physics that has to do with heat transfer.

Joseph Fourier was a French engineer who studied how metals cooled. His main objective was to make more efficient cannons for Napoleon's armies. In so doing, he discovered what is now known as Fourier's law of heat conduction. This law links how much a substance is heated up in a given time with how much thermal energy is emitted across its surface. The details can be found in Appendix 4, but, in summary, Fourier's law implies that if the ice is heated up in a time t, the depth to which the heat will penetrate is roughly $d \approx \sqrt{\frac{\kappa t}{\rho C}}$, where $t = L/v$, the time it takes for the blade to pass over a point on the ice. The constant κ is the thermal conductivity for ice, defined as the rate of heat in J/s passing through a body one meter thick with a cross section of 1 m² when the temperature difference on either side is 1°C. For ice, $\kappa = 2.1$ Jm/s/°C.

Putting all this together gives:

$$\Delta T \approx \frac{1}{2} \frac{\mu m g}{A} \sqrt{\frac{Lv}{\rho \kappa C}}. \tag{1.4}$$

This equation suggests that the greater the velocity, the greater the temperature rise will be. This does not mean that more heat is created per unit of surface; rather, heat is dumped more quickly into the ice and therefore has less time to dissipate, making the top layer hotter.

To get an estimate of a typical ΔT we use the following parameter values: $\rho = 920$ kg/m³ (ice is less dense than water), $C = 2,220$ J/kg/°C (ice has a lower heat capacity than water), $v = 5$ m/s, $m = 80$ kg, $\mu = 0.005$ (typical, as we will see in the next section), $A = 6$ cm², and $L = 10$ cm. Plugging these into Equation 1.4 yields a temperature rise of 2.2°C, which is quite a bit more important than the effect of pressure melting.

Of course, this back-of-the-envelope calculation is not meant to be accurate, even though it compares well with more detailed analysis.[6] It does not take into account, among other things, the latent heat of fusion needed to melt ice and all the dissipation of heat through the ice and the blade. The equation is therefore only an upper limit to the temperature rise, but it makes the point that frictional heating is a more potent means of melting ice than pressure.

Although frictional melting may be a factor under a skate—easy to believe when looking at red-hot skaters such as Anaheim's Paul

---

6. Ibid.

Kariya, who seem to leave a trail of vapor behind them—it does not explain everything. Even a temperature rise of a few degrees is not enough to reach the melting point in all cases. Don't forget, if the ice is kept at −10°C, a rise in temperature of 3°C means we're still at −7°C. That being said, arguments for frictional melting are supported by the fact that the coefficient of friction drops as skating velocity increases. Thus it still plays a secondary role.

So how is it possible to skate on very, very cold ice, say at −200°C? (Not that anyone has ever tried! Yet we do know, as an experimental fact, that ice is slippery even when that cold.) How can we explain sliding when very little pressure or velocity is involved? For example, light objects like pucks slide just as well as skaters, yet the frictional heat generated is minute. For a slap shot at 100 mph, our formula predicts that a puck with a mass of 170 g creates a temperature rise of only 0.002°C! There must be another factor at play, so we turn to another process to solve the mystery.

3. The Quasi-fluid Water Layer

Today's physicists, armed with lasers, atomic microscopes, and other highly sophisticated equipment, have recently tackled the question of ice's slipperiness head on, and the results have been interesting. They found that regardless of pressure or friction melting, the surface *itself* is slippery. In other words, you don't need to melt ice at all in order to slip on it: down to temperatures as low as −250°C the surface has a wet layer of quasi-fluid water. The layer is very thin, but it is enough to provide lubrication. It has a consistency similar to that of slush, and, as the temperature rises, the thickness of this layer increases and the ice surface becomes more slippery.

To understand this further, we need to examine ice at the molecular level again. First, the molecules at the surface are not attached to the rest of the ice with the same strength as molecules packed inside. Because the top molecules don't have any "upstairs neighbors," they have unfilled bonds that reach up, ready to grip at anything passing by. These molecules are therefore plucked away rather easily. Even when little pressure is applied, like when your foot slips on the ice, the top molecules are easily detached and "roll" under your shoe like tiny ball bearings. They remain in an almost liquid, slushy state.

Another reason for the slipperiness of the ice surface has to do with the crystal geometry of ice. The layered structure discussed earlier facilitates the sliding and detachment of thin ice sheets. Because the layers are not tightly bound to one another, a small shear force— a sideways force along the surface, like that caused by a skate— will make them slide on top of one another. In this regard, ice is similar to another common solid, graphite (the material incorrectly referred to as the "lead" in a pencil). Graphite is a form of carbon, and, for those who remember such things, it was the sort used to make carbon paper. Technology may have surpassed this tool, but the terminology—"cc," for "carbon copy"—has lasted into the age of the Internet. Graphite is a dark material that is quite soft and brittle, and its layered structure makes it suitable for writing. When rubbed against paper, the shear force detaches graphite layers and leaves a thin, very visible trace: the pencil line. Similarly, when they come in contact with a moving object, small sheets of water molecules detach and move across the ice surface, acting like the rollers used at airport baggage conveyors.

A recent study done by a team of scientists at the University of California at Berkeley probed the ice surface with a high-tech microscope.[7] The so-called "atomic force microscope" can give images with a resolution at the atomic level. Technology that allows scientists to actually see individual atoms and molecules was developed about two decades ago and earned the inventors a Nobel Prize in physics in 1986. However, unlike conventional optical microscopes —the ones high school students use to see bugs up close—the atomic microscope "feels" rather than "looks" at a specimen. It has an extremely small, sharp tip of only a few atoms across. This tip is brought close to a surface and moved around, giving the microscope information about the surface structure and its roughness as well as the lateral force acting on the tip. As seen earlier, the lateral-force information is used to determine the friction coefficient of the material.

Applying this technology to ice revealed two important pieces of information. By scanning a tiny ice crystal, researchers determined that even at a temperature of $-24°C$ the surface had a quasi-liquid

7. H. Bluhm, T. Inoue, and M. Salmeron, "Friction of Ice Measured Using Lateral Force Microscopy," *Physical Review B* 61 (2000): 7760.

layer that was approximately 8 nanometers in thickness (1 nanometer = 1 millionth of a millimeter). This is equivalent to about 10 times the diameter of a water molecule. There was a surprise in store, though. Because the tip of the microscope is so small, it could be dipped right through the top, quasi-liquid layer and make contact with the solid ice surface to obtain a "dry" coefficient of friction for ice. This turns out to be a magnitude of 0.6, similar to common solids such as plastic, rubber, and metals. So, were it not for ice's "wet" layer, playing hockey on ice would be just as hard as trying to skate on a sheet of plastic!

Any Conclusion?

Having considered pressure, frictional heating, and quasi-fluid water, how can we sum it all up? Which one of these effects is most important in making our favorite NHL athletes move so fast? Skating is possible at common rink temperatures because of the layer of quasi-fluid water that exists even on a bare ice surface. Pressure and friction are merely mechanisms that help lower the friction coefficient by melting the ice, although the effect of pressure is practically negligible. On the other hand, frictional heating can be observed indirectly by measuring the friction coefficient as a function of speed, as we shall see shortly.

Friction Force Measurements

As far as hockey is concerned, the magnitude of the coefficient of friction is more important than the reasons for slipperiness. Various researchers around the world have experimentally measured friction between the blade and the ice during skating. A low friction coefficient is particularly important for speed skating competitions, as world records may stand or fall depending on ice conditions. During skating, the dynamic friction coefficient is the one of interest, because the blade is technically always moving.

To measure the rubbing force, scientists installed sensors between the blade support and the boot. This way, both the horizontal forces (along the ice) and vertical forces (along the leg) could be directly measured. These components are proportional to the friction and push-off forces, respectively. Taking the ratio of the horizontal and the vertical components, the friction coefficient is obtained. With this technique, scientists have studied the dynamics of skating under various ice temperatures and skating velocities.

Fig. 1.7 shows the results for the friction coefficient at different ice temperatures for an athlete skating at 8 m/s.[8] All measurements were taken on the same ice rink, and ice temperature was changed over a period of several hours. The data are quite scattered, partly owing to the unavoidable changes in the skater's pace and pushing technique, but it appears that friction is lowest from approximately −9°C to −5°C. As predicted by the quasi-liquid layer theory, the coefficient decreases steadily with increasing temperature, up until

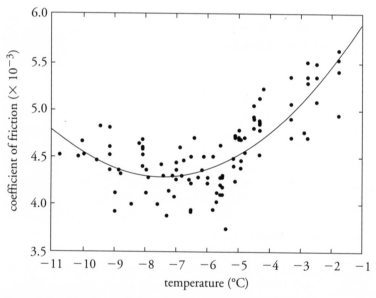

Figure 1.7. Friction coefficient of ice as a function of temperature, measured directly on an athlete skating at a constant velocity of 8 m/s.

8. J. J. de Koning, G. de Groot, and G. J. van Ingen Schenau, "Ice Friction during Speed Skating," *Journal of Biomechanics* 25 (1992): 565.

−7.5°C. Above that, friction increases again. Why is this? As the ice becomes warmer and softer, the blade cuts a deeper groove into it, and this larger groove means more resistance and more lost energy. In fact, measurements show that ice friction is at its highest point when the blade first contacts the ice and digs into it. After that point, the skate glides smoothly and friction is lower, but it rises again at the end of the push when the skate rotates and makes another groove. When hockey players refer to "hard" ice as being fast, it is this mechanism that is at play. So the shape of the graph in Fig. 1.7 is representative of a trade-off between slipperiness and hardness. The optimal point in this case lies around −8°C to −7°C.

The influence of velocity on the friction coefficient is somewhat controversial. As early as the 1950s, F. P. Bowden and others showed that the friction force between skis on a sled and ice decreased with velocity, which supports the idea that frictional heating helps create a lubricating film.[9] On the other hand, J. J. de Koning and his colleagues found that friction increased with speed when measured directly on the athlete's skate.[10] In their case, the coefficient ranged from 0.004 at 5 m/s to 0.006 at 10 m/s. But these seemingly conflicting results may both be right in their own way! The difference is that in the first study, the sled keeps the same contact with the ice at all velocities, whereas in the second one the skating technique of the athlete likely changes throughout. At higher speeds, the skater must bend lower and push harder to overcome the wind. This results in deeper ice deformations, which explains the greater friction coefficient. Another important factor is the temperature of the ice itself. When ice is very cold, warming it up with frictional heating might make it more slippery, as Fig. 1.7 suggests. Yet the opposite trend occurs when the temperature is closer to zero. This idea is supported by experiments that found the relation between velocity and the friction of ice on ice was strongly dependent on the temperature.[11] So, although ice

---

9. F. P. Bowden, "Friction on Snow and Ice," *Proc. Roy. Soc. A* 217 (1953): 462.

10. De Koning, de Groot, and van Ingen Schenau, "Ice Friction during Speed Skating."

11. G. Casassa, H. Narita, and N. Maeno, "Shear Cell Experiments of Snow and Ice Friction," *Journal of Applied Physiology* 69 (1991): 3745.

friction alone decreases with speed in the case of uniform contact, the total resistive force on a skater can still increase, owing to the change in the skating technique and greater ice deformation.

To make matters even more complicated, two ice surfaces can have different amounts of slipperiness even under the same conditions of temperature and other environmental parameters. The reason for this variance lies in the surface chemistry, which plays a role in determining the consistency and the thickness of the wet layer. Hence, impurities and chemical agents can turn "fast" ice into "slow" ice, or vice versa. For instance, it was reported that adding a small quantity of ethylene glycol could yield a 30 percent better friction coefficient.[12] Icemakers at some speed skating rinks add "secret" chemicals—usually surface-active agents, known in the trade as surfactants—to give the ice a superior quality. Water purification systems are sometimes used, and conditioning agents are added to remove unwanted chemicals present in tap water, such as alkaline salts, which tend to reduce slipperiness and dull the skate blades. Surfactants, a common ingredient in soaps, are molecules that have a water-loving end and an oil-loving end. Normally, oil and water don't mix. A surfactant can hook a water molecule with an oil molecule, however, and the whole unit floats on water. The result, in the case of dishwashing, is that you end up with clean dishes and some nasty-looking scum floating on the surface. Environmentalists use surfactants to break up and remove oil spills. Surfactant molecules also allow icemakers to get rid of unwanted chemicals that could ruin the ice for hockey players.

Ice-skating, along with bobsledding, is among the ultimate low-friction sports. Rollerbladers have to put up with friction five times higher than their ice-cruising counterparts, while skiing on snow is 10 times more resistive than ice-skating.[13] Only cycling has a comparably low coefficient of friction, thanks to the large wheels that help minimize the energy loss between the tires and the road. Anyone who has seen a professional NHL game understands that the hockey

---

12. De Koning, de Groot, and van Ingen Schenau, "Ice Friction during Speed Skating."

13. P. E. Di Prampero, G. Cortilli, P. Morgnoni, and F. Saibene, *Journal of Applied Physiology* 47 (1979): 201.

players benefit from this low friction—as they can put all that saved energy into antagonizing each other!

Blame the Ice

Hockey players and commentators alike sometimes complain about improper ice quality. Figure skaters are also known to blame the ice for a poor performance. Whether every such claim is justified or not, it's true that not all ice surfaces are created equal, even within the NHL circuit. For instance, the ice in Edmonton is often cited as fast and high-quality, whereas at other rinks the puck sometimes seems to bounce around like a ball. Hockey fans may remember a freak goal caused by a bad puck bounce during one game or another. We also hear about "fast ice" and "slow ice" (though less often than we hear about "fast players" and "slow players").

Even in the course of a game the ice quality can change, and this affects the way the game is played. As snow builds up, the puck sticks more to the ice, and so players will tend to be cautious and not stickhandle as much. They will tend to be conservative rather than use finesse. The increased coefficient of friction also affects skating. Because of scratches and bumps, skating with the same speed requires more energy toward the end of a period.

Left on its own, ice normally becomes quite rough. Without resurfacing and polishing, irregularities eventually appear all over. Outdoor ice can be particularly bad. Growing up in the province of New Brunswick, my friends and I often played hockey outside on frozen puddles after school. We would play until 4 or 5pm, when darkness forced us to quit. Often, the ice was so rough that skating would feel like getting a vibrating foot massage, especially if the pond happened to have frozen on a windy day.

Making a fine sheet of ice is almost an art. Each arena has its particular geometry, environment, and refrigeration apparatus, so the tricks to be learned vary from one place to another. In new arenas, bringing the ice under control can take a while, usually a few months. In February of 1999, the Toronto Maple Leafs moved out of the good old Maple Leaf Garden into the new Air Canada Center. At first, everything in the more spacious and modern building seemed perfect, but the ice soon proved inferior—despite the fact that it had been branded as a state-of-the art playing surface. Players complained about the choppiness and inconsistency of the ice during the first few games. The team, led by captain Mats Sundin, was built for speed, so playing on a slower, rougher surface hurt their game: "home-ice advantage" had become "home-ice disadvantage." The proper adjustments were eventually made and things returned to normal, but it took months before the players stopped grumbling.

Chapter 2

SKATING

The most basic skill in hockey is obviously the ability to move around on the ice quickly and efficiently. During a race for a loose puck or a breakaway scoring chance, a player like Maurice Richard, with great speed and agility, always has an edge over his competitors. Avoiding body checks (or boarding) and reacting quickly to a play demand good skating skills. Although not all good skaters are good hockey players, it's safe to say that every good hockey player is, above all, a good skater. It comes as no surprise that the fastest skaters in the NHL—those players who compete in the sprint event during the All Star competitions, like Pavel Bure of the Florida Panthers and Peter Bondra of the Washington Capitals—are also among the top scorers.

When watching the effortless grace of Paul Kariya as he sprints across the ice at the Arrowhead Pond in Anaheim, it might be hard to remember that, at least from a scientist's point of view, skating is a very complex set of actions. The science and biomechanics of skating have already been extensively studied, not so much in the realm of hockey but rather for speed skating, a winter sport in which every millisecond counts. Understanding the physics and the underlying mechanics can help coaches and players improve their skating technique. There are, of course, some major differences between speed skating and hockey, such as the shape of the blade and clothing, but the basic principles are the same, as is the ultimate goal: greater speed, acceleration, and agility. To start things off, let's take a look at the most important piece of equipment, the skate.

The Skate

The skate is a uniquely designed piece of footgear that was only recently invented. Skis, in contrast, appeared thousands of years ago, and snowshoes could even be older. Archeologists uncovered ski fragments and pictographs of skiers in Norway dating from at least 2000 B.C. We know that the Carthaginian soldiers and their leader, the implacable Hannibal, faced avalanche hazards in the Alps on their way to attack Rome during the war of 218–201 B.C. It is almost certain that they used skis or snowshoes to cross the dangerous snowy mountains.[1]

Early skates, meanwhile, began to resemble the ones we have today only around the mid-nineteenth century. An American named E. W. Bushnell invented a skate with a steel blade in 1850, a date that, not surprisingly, roughly coincides with the beginnings of hockey. Prior to that, no one used a specialized shoe to move on ice—you simply strapped some type of contraption, usually made of metal or wood, to the bottom of your regular footwear. Over the ages, skating—or, more accurately perhaps, *gliding*—had become a standard mode of locomotion in Nordic countries with long winter seasons. The earliest skates were discovered in Sweden and date back to the ninth century. Primitive skates dating from the same period were also uncovered at Viking settlements in Britain, giving Scandinavia a legitimate claim as the birthplace of skating.[2] However, these ancient artifacts have little in common with the Bauer and CCM skates worn by today's hockey players: the "blades" were made of ground and flattened bones from the foreleg of a reindeer or a cow.

The modern hockey skate has a number of important characteristics worth mentioning. First, the rounded shape at the front and the back of the stainless steel blade permits greater skating flexibility. Players often need to make sharp turns or lean forward to grab a puck. This would be awkward if the blade were flat from end to end. During a turn, the skate is inclined to its side, so a rounded blade helps the skater follow a circular path. Flat blades, like those on speed

---

1. D. Lind and S. P. Sanders, *The Physics of Skiing* (New York: Springer-Verlag, 1996).

2. P. J. Vesilind, "In Search of Vikings," *National Geographic* 197, no. 5 (2000).

skates, tend to keep a straight line. (To convince yourself, try to slide and rotate a plastic credit card on a carpet. Then do the same with a large coin.) This is why speed skaters must constantly shift their feet to the inside when they are on a curved stretch. Flat blades are not appropriate for figure skating either: Elvis Stojko could never spin or perform his triple-axels without slightly rounded blades.

The blade is firmly attached to the sole of the boot with plastic or metallic fixtures. The boot itself is usually made of leather or, more commonly these days, nylon and other hardy synthetic materials. Engineers in the skating industry are continually looking to improve the design by using materials that are more resistant, durable, and comfortable to wear. Recently, special gel-like substances were introduced in the inside padding of the boot; this gel molds to the shape of the ankle and provides a better grip on the foot.

The boot, as every hockey player knows, is very stiff and has a hard toe area to protect from the impact of an oncoming puck. Players need to lace their boots up tightly above the ankle in order to prevent ankle motion, especially lateral (side-to-side) movement, which may cause injuries. Loose skates put an unnecessary amount of stress on the ankle and create instability. The ankle shouldn't be allowed to move sideways in order for the leg to push consistently with each stroke. However, the boot permits a limited amount of medial (front-back) ankle motion, to allow the calf muscle to push forward and to ease knee-bending when the player crouches.

One drawback to a stiff and tightly laced skate is that it tends to cut off blood circulation to the foot—and without blood flow, the body's extremities cannot keep warm. Sometimes kids learn this the hard way when they unlace at the end of the game on a cold day. As the normal blood flow returns to the feet, it is a very painful experience to realize that your feet have been freezing the whole game!

It's All in the Groove

Many years ago, during a pregame warmup, I once saw a teammate fall flat on his face as he stepped onto the ice: he had forgotten to remove his plastic blade protectors. Without a grooved blade making contact with the ice, he was unable to move around or keep his balance. I should have learned from his experience, but I made a similar mistake later. When asked if I wanted my skates sharpened with a regular cut, a one-sided cut (sharper on one side than the other), or a flat cut (almost no groove), I opted for the last one. I'm a goaltender, and I figured a flat cut would make sideways motion easier, helping me cover all sides of the net. I soon discovered a flat cut was a big mistake, for I found myself unable to move in any direction at all! I ran back to the shop and asked for a regular cut, then used a piece of sandpaper (probably not the best solution since it can damage the steel blade) to dull the inner side of the blade for easier sideways motion.

So, through this unintended experiment, I learned that the hollow grinding of the skate plays a crucial role. Figure 2.1 shows an enlarged cross section of the blade and the arc-shaped groove along the blade axis. This groove is necessary to stabilize the skate in the lateral direction. Because of the small area underneath the blade and the large pressures exerted upon it (see the previous chapter), the sharp edges easily penetrate the ice, providing enough grip for a skater to stop on a dime, kick up a spray of ice, or accelerate quickly. Another way to look at it is to consider the skate as having a highly asymmetric coefficient of friction: it offers little resistance to forward or backward motion but a lot of resistance to the side. In fact, the ratio between the two coefficients of friction may be as high as 200 or more. With no groove, the ratio would be close to 1. If a player needs the ability to move sideways frequently (as goaltenders do), skates with less-pronounced grooves are preferable. But in all cases, the frictional asymmetry of

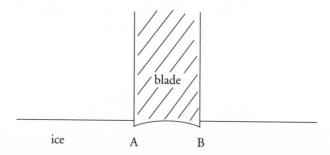

Figure 2.1. Enlarged cross section of the skate blade in contact with the ice surface. Points A and B are where the greatest pressure and deepest penetration occurs. Lateral motion is restricted by the groove.

the skate provides the grip needed to counteract the slipperiness of the ice. Without the groove, you might as well be barefoot.

The scientific question that follows is whether there is an "ultimate" groove geometry, one that works better than the conventional, arc-shaped one. For example, how about a blade with a single, triangular central edge like a thick knife? How would that work? Because sharpening a blade into this kind of edge is easier than making a groove, chances are that skate makers experimented with such designs early on, but they would have abandoned them. A single edge is inefficient because the blade would dig too deeply into the ice, making the friction larger than necessary. A greater deformation of ice means more energy is wasted during gliding. The wide, hollow groove used today is probably close to the ideal geometry, as it prevents lateral motion on both sides while not sinking too deeply into the ice.

Goalie Skates

When I was about eight years old, I used to play "atom-level" hockey at my hometown minor league in St.-Isidore, New Brunswick. It was quite a scene: a bunch of kids wearing purple jerseys, struggling to get organized on the ice and put on a good show before the small crowd. At one point the coach began looking for a regular goalie, so he made us take turns guarding the net. We would use the arena's old goalie equipment, each piece of which always seemed too big or too small. When my turn came, I discovered that goaltending was kind of fun, so the coach and I decided I would become the team's netkeeper. For a while I kept my regular skates on, but it was awkward and I seemed to be constantly falling forward or backward. At the time, I thought nothing of it, but when I got older and became more serious about stopping pucks, my dad purchased a pair of real CCM goalie skates along with a set of equipment that fit—a wise choice to avoid injuries. Indeed, one of the special characteristics of the goalie skate is the hard protective shell, to protect against pucks. Although I didn't like the look and feel of my new skates at first, they eventually made a world of difference. For one thing, their flatter, longer, and slightly wider blades gave me more balance, so I could focus on stopping pucks rather than on staying upright. Sideways motion was also easier, which is crucial in a five-on-four power play or in a two-on-one break when there's a lot of quick, cross-ice passing in front of the net.

Goaltenders often need to kneel down to block a shot and rise quickly to be ready for the next. Being caught off balance is the number-one danger. Colorado's Patrick Roy could never have turned his famous "butterfly" technique into an effective goaltending style without the aid of proper goalie skates. There is one trade-off to using goalie skates: it is harder to turn with a long, flat blade. But turning in tight circles is not as critical for goalies as for regular players.

Another limitation of the goalie skate, and all skates for that matter, is that the grip is lost when the blade is off the ice. To prevent this from happening, someone had the clever idea of attaching a small piece of sharpened metal under the sole of the boot, beneath the big toe. This way, when the main blade leaves the ice, the small one keeps contact, aiding mobility. The extra blade is particularly useful to goaltenders who are often on their knees and use the butterfly technique. According to its inventor, Toronto resident John McLeod, the modified blade also helps reduce the risk of knee injuries.[3] Although Ottawa Senator Patrick Lalime had the longest undefeated streak by a rookie goaltender while using them, the skates didn't gain wide acceptance and were even banned by the NHL in 2001.

Why Not Speed Skates for Hockey?

In the late 1990s, world records in speed skating were suddenly shattered on virtually all distances. Race times fell by as much as 4 percent at once, a huge amount for a sport in which improvements are usually measured in hundredths of a second. The reason was the arrival of a new kind of skate, the klapskate. Some skaters who had been barely known before became acclaimed record holders because of their wise decision to make the switch from conventional skates.

The klapskate is a simple but effective improvement over the conventional speed skate. It has a hinge on the front fixture of the blade and there is no attachment at the back, so the boot can tilt forward while the blade remains in full contact with the ice. The new design

3. D. Edwards, "Dryden Backs Ban on Goalie Device," *Globe and Mail* (Toronto edition), August 9, 2001, p. S2.

enables the leg to provide a longer push without increasing the ice friction on the blade. It didn't take long before puzzled sport scientists took a close look at the klapskate to see how it affected skating. They found that the added foot flexibility allows the leg to generate more power not only at the ankle but also at the hip. This simple improvement increased the athlete's power output by as much as 10 percent.[4]

One may wonder why hockey hasn't made use of this speed skating technology. Wouldn't players become faster and therefore better? While it is true that, with some training, athletes can attain greater speed on a linear stretch with speed skates, we should not forget that hockey is as much about mobility as it is about speed. As mentioned earlier, the unavoidable sharp turns and quick responses to a change of play make the rounded blade a necessity. We will see in the next chapter that when a player makes a slap shot, there is a transfer of body weight from the back leg to the front, in the direction of the shot. This is accompanied by a quick rotation of the skates. Without the quick foot positioning made possible by the rounded blade, such a move would be awkward.

It is not yet certain whether new design features such as the klapskate will eventually appear in the hockey rink, but it is almost certain that a hinged blade would not have the impact it had in speed skating. For one thing, because the hockey blade is not flat, it would not stay parallel to the ice as the boot is tilted forward, thereby defeating the purpose of the design.

The Biomechanics of Skating

A professional ice skater gliding over the rink can be a graceful and beautiful sight. For a scientist, though, ice-skating is less prosaic: it is a sequence of leg and trunk movements repeated at regular intervals. Every cycle of motion—also called a *stroke* or *stride*—involves each leg taking a turn at propelling the skater in a given direction, an action called the *push-off.* It's quite a workout, involving four limb segments

---

4. H. Houdjik, J. J. de Koning, G. de Groot, M. Bobbert, and G. J. van Ingen Schenau, "Push-Off Mechanics in Speed Skating with Conventional Skates and Klapskates," *Medicine and Science in Sports and Exercise* 32 (2000): 635.

Figure 2.2. We see on Sami Kapanen the four semirigid body segments involved during skating: the foot (AB), the lower leg, or tibia (BC), the upper leg, or femur (CD), and the torso (DE). CP Picture Archive (Kevin Frayer).

and three major skeletal joints (see Fig. 2.2). The major body parts used are the foot, the lower leg, the upper leg, and the torso, connected via the ankle, the knee, and the hip joints, respectively. Most of the skating motion comes from flexing these three joint groups. Most of the energy needed to propel the skater is released by the calf muscles (with ankle flexion), the quadriceps (with knee flexion), and the hips and buttocks (with hip flexion). Some parts of the upper body are also involved in keeping a steady course and delivering more energy. By swinging back and forth, shoulders and torso help the legs to deliver the maximum force and energy.

In spite of involving all these muscle groups, skating remains one of the most efficient human-powered modes of locomotion, thanks, as seen earlier, to the low friction of the ice. According to sport scientists, a person weighing 175 pounds will burn roughly 135 calories per mile running but only 65 calories per mile skating—roughly half the amount. This compares well with the 50 calories needed to cycle a mile (cycling being the most efficient way of going from point A to point B).

Unlike running, in which each leg pushes forward in a similar fashion regardless of speed, skating involves more complex movements because the asymmetric nature of the skate imposes restrictions on the way motion is achieved. On top of this, an athlete has to master several skating techniques to become an accomplished hockey player: linear (straight line) skating to pierce the offensive zone, circular skating to go around the net, backward skating, and of course braking, the opposite of skating. These modes of skating each have their own complex mechanics. Though hockey only dates back to the mid-1800s, the physics needed to describe it have been around a bit longer. Isaac Newton, who laid out the framework for the science of mechanics in the seventeenth century, would have understood the physics involved.

The Race for the Puck: Moving Forward

The simplest mode of skating is straight, linear motion. Because ice is almost without friction, you can't really speed up by "running" with your feet pointed straight ahead—there's nothing to push against. As Newton put it, action and reaction are equal and opposite, so without friction to cause a reaction (that is, to push your body forward), you're unable move. This means that hockey players must propel themselves with a series of sideways pushes, as Fig. 2.3 illustrates, with each blade sticking out at an angle θ relative to the forward

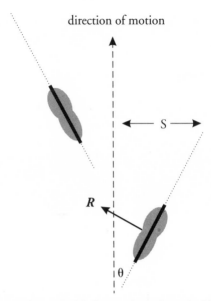

Figure 2.3. Due to the negligible friction force along the blade, force must be applied perpendicular to it. The width *s* is the lateral displacement of the skate during the push-off.

direction. During a stroke, one blade makes contact with the ice for the push-off while the other one is raised and moved ahead before gliding on the ice. Sometimes the gliding skate is oriented straight ahead at the beginning, then turns toward the outside when it starts to push. Evidence of this can be seen from the curved marks left on the ice.

Pushing sideways on the ice is done in one of two ways: by staying upright and moving the legs to the side (you don't need to bend your knees at all), using mainly the hip muscles; or, as is done to provide powerful acceleration in hockey and speed skating competitions, by inclining the body forward and pushing with the leg, using the more powerful quadriceps (see Fig. 2.4). In both cases, the ice reacts with a force that has a small component in the forward direction. This component is the only one that matters as far as forward propulsion is concerned. (Readers who are not familiar with the concept of a force component should refer to Appendix 3 for a discussion on how

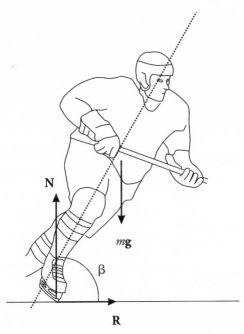

Figure 2.4. Leaning forward allows the leg to exert a greater amount of force. When a hockey player accelerates (or fights against air drag), his center of gravity (mg) moves ahead of the point of contact with the ice (where N is applied).

to manipulate vector quantities.) Because ice friction is very small, the reaction force has virtually no component along the blade axis, as Fig. 2.3 shows, but is instead oriented perpendicular to it. Since the blade is oriented at an angle θ, trigonometry tells us that the forward component of the force is $R \sin \theta$, where R is the reaction force along the surface of the ice (the vertical component doesn't contribute to horizontal motion).

In the simplest case, when you are skating with the help of your hip muscles without bending your knees, most of the push-off force is oriented along the ice, and R is simply that pushing force. On the other hand, when you propel yourself by bending your knees and leaning forward, the push-off force has a large vertical and horizontal component. If we suppose you're pushing with a force F from your leg and you're leaning at an angle β relative to the ice ($\beta = 90°$ if you are standing upright), then the reaction force along the ice is $R = F \cos \beta$, and the forward component of that force becomes $R = F \cos \beta \sin \theta$. Now that we know the force in the forward direction, we can deduce the acceleration from Newton's second law: it is $a = R/m$, where m is your mass.

These equations relate the skating acceleration to the push-off force and the orientation of the skates. What is to be learned here? How can we use this to improve our skating? First of all, if your skates are straight ($\theta = 0°$), you will not move forward, only side-to-side. The greatest acceleration is achieved at the largest angle θ. When speed skaters or hockey players want to accelerate quickly from rest, their stakes are really sticking out to the left and right. The second lesson is that leaning forward (keeping β small) is very important. This is why speed skaters bend forward at the beginning of a race and hockey players crouch forward as they try to gain speed. Last, the push-off force is proportional to, well, the force you push with! Therefore, you need strong legs to accelerate quickly.

The push-off and the reaction forces are not constant in time, of course, but change substantially within the stride. Scientists have measured this force experimentally using sensitive electronic devices sandwiched between the blade and the skate boot.[5] Results of their experiments are graphed in Fig. 2.5. It shows a typical push-off force on one skate over a period of time totaling four strokes. Notice that the force is largest at the beginning of each stroke, when the sharp edge of the blade penetrates the ice, and at the end, when the leg is almost fully extended and can exert a greater force.

But is pushing side-to-side the only way to move on ice? Not really. Figure skates, for example, have jagged picks at the front that can grip the ice and allow a skater to move ahead by a walking-type motion. But pushing sideways is the technique that produces the greatest velocities, allowing skaters to reach speeds even faster than they can move their feet—estimated to be around 7 m/s.[6] The fastest

5. J. J. de Koning, G. de Groot, and G. J. van Ingen Schenau, "Ice Friction during Speed Skating," *Journal of Biomechanics* 25 (1992): 565.

6. G. J. van Ingen Schenau, R. W. de Boer, and G. de Groot, "On the Technique of Speed Skating," *International Journal of Sport Biomechanics* 3 (1987): 419–31.

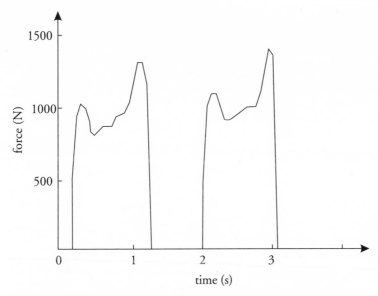

Figure 2.5. The push-off force on the blade as measured directly on a skater cruising at a constant speed of 8 m/s. From J. J. de Koning, G. de Groot, and G. J. van Ingen Schenau, "Ice Friction during Speed Skating," *Journal of Biomechanics* 25 (1992): 565.

runners, for instance, move their feet at 12 m/s. So it is impossible for a skater to push against a fixed point on the ice if he or she is moving faster than that. Hence, speeds greater than this limit would not be possible without the push-and-glide technique. Even though the side forces cause the body's center of gravity to follow a sinuous trajectory (with an amplitude observed to vary between 25 and 50 cm), the gliding technique enables skaters to reach speeds exceeding 50 km/h, far more than what could be achieved through simple running.

As mentioned earlier, when a skater starts from rest, the skates are oriented at a very large angle θ in order to reach maximum acceleration—in other words, the skates make a V shape. But as a skater picks up speed, this angle must be reduced if the stroke period (the time needed to accomplish one stroke) is to remain the same. An easy estimate of the angle as a function of speed can be obtained if the stroke period T and the displacement s of the skate are known (see Fig. 2.3). Using trigonometry, we find that the sideways velocity v_s of the skate and the skater's velocity v are related via $v_s/v = \tan \theta$. Since $T/2 = s/v_s$, then

$$\theta = \arctan \frac{2s}{vT}. \tag{2.1}$$

For example, assuming $v = 6$ m/s, $T = 1$ s, and $s = 60$ cm, the result is $\theta = 11°$.

At constant velocity, each leg takes a turn pushing and gliding to overcome the resistance of ice and air. Much of hockey is accelerating and slowing down quickly, and when explosive acceleration is needed, a great deal of pressure is applied to the lower part of the body. In technical language, the leaning angle of the body axis—roughly defined as the line crossing the center of the players' mass and the point of contact on the ice (see Fig. 2.4)—relative to the ice is less than 90°. The immediate question that occurs to a scientist is to ask how this leaning angle corresponds to a given acceleration. To answer this, we need to simplify greatly and assume that the player is a complete stiff—that is, he is a rigid body. When the skater accelerates forward, his center of mass stays pretty much at a constant height. Therefore, the sum of all vertical forces acting on the player's center of mass must vanish. If we call N the vertical (normal) and R the horizontal reaction forces of the ice, we have the following equation for the force applied at the center of mass:

$$(N \sin \beta + R \cos \beta) \sin \beta - mg = 0. \tag{2.2}$$

Because N directly counteracts the body weight, $N = mg$, and, since R causes the forward acceleration, we have $F = ma$. Putting these two pieces together reduces Equation 2.2 into a nice little formula:

$$\tan \beta = \frac{g}{a}. \tag{2.3}$$

This tells us that the greater the acceleration, the smaller the angle β is, and if $a = 0$ then β must be 90°—that is, the player is standing up straight. Starting from rest, a typical hockey player is able to accelerate at a rate of about $5m/s^2$, corresponding to a leaning angle of 60°. Interestingly, this angle does not depend on the player's height or weight but only on the acceleration and the gravitational constant g. On the moon, where g is six times smaller than on Earth, hockey players could lean much lower, down to about 20° from the horizontal (assuming the skates are sharp enough to grip the ice)—but an NHL franchise has yet to be sold there!

The total force F exerted by the leg is obtained by adding vectors N and F:

$$F = \sqrt{(mg)^2 + (ma)^2} = mg\sqrt{1 + (a/g)^2}. \qquad (2.4)$$

To take a concrete example, a player accelerating at 6 m/s^2 is exerting a force equivalent to roughly 1.17 times his own body weight.

Powerful accelerations don't last more than a few seconds. If you keep track of one particular player during a hockey game, you will find that he typically makes a sequence of three or four accelerating strokes at a time, then keeps a constant speed, turns, or brakes. As speed picks up, acceleration decreases, the velocity converges toward the maximum level, and strokes return to the steady-speed mode.

Another type of linear skating is backward skating, often practiced by defensemen who need to move toward their own net while facing their opponents. One of the most accomplished skating defensemen was Bobbie Orr, who has appeared on many NHL All Star teams. Looking at him play, it seemed like he could skate backward and forward with the same ease and grace.

Players on defense usually don't start from a resting point moving backward. Instead, they skate forward toward their zone and then make a 180° turn near the center-ice. The reason is simple: it's quicker. Having accelerated, once spun-around the defender only needs to keep a constant speed. If one really wants to skate backward from rest, it is possible to do so with a stroke sequence like the one depicted in Fig. 2.6. Each skate follows an S-curve and takes turns pushing backward and gliding, all without leaving the ice. The acceleration achieved this way is not nearly as great as when skating forward, partly because the hip muscle can't contribute as much force.

direction of motion

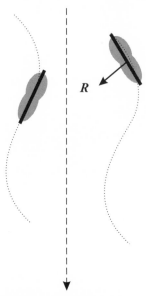

Figure 2.6. Backward skating is accomplished with a wavy motion. The pushing skate is pointed inward and transmits the reaction force of the ice to the body, while the other skate glides.

Going Nowhere: Circular Skating

Hockey is a game of close and intense interaction, in which players are forced to turn constantly. With a dozen aggressive skaters on a rink measuring only 200 feet long and 85 feet wide, changing direction is unavoidable—it's part of what makes hockey exciting. Who can forget the quick turns and wiggles of the young Wayne Gretzky as he baffled his opponents on his way to the net?

The question is, How do we turn in the most energy-efficient way? Saving energy and momentum is important if you are going to last through the game. After all, in a closely fought battle such as Game Four of the Flyers-Penguins Stanley Cup semifinals in May of 2000, which lasted into a sixth overtime period, it helps to have expended as little energy as possible during regulation time. The best strategy to change direction, it so happens, is to move in a circle with a short radius.

The mechanics of circular motion at constant speed is a staple of introductory physics courses. It is relevant not only to sports but to everything that turns, from wheels to propellers to planets. Circular motion is simply described in terms of an acceleration directed toward the center of the circle (hence its name, *centripetal*) resulting from the continual change in the direction of the velocity. According to Newton's first law, which states that an object will stay at rest or continue moving in a straight line unless a force is applied to it, there must be a force involved. When you spin around while holding a bucket of water, your hand provides the centripetal force needed to keep the bucket in a circular path. When you let go, the force vanishes and the bucket momentarily goes in a straight line. Likewise, a hockey player turning a corner must rely on a force to keep to a circular path; this centripetal force comes from his skates gripping to the ice.

The inward acceleration of a body moving in a circle is given by

$$a = \frac{v^2}{R}, \tag{2.5}$$

where R is the radius of the circular path and v is the velocity. Therefore, the tighter the radius, the greater the acceleration and the larger the force needs to be. Just as with linear acceleration, a skater needs to shift his or her center of gravity and lean inward and beside the point of contact on the ice. The faster the speed, the smaller the angle of this lean (see Fig. 2.7). The angle of the body axis relative to the ice is calculated by combining Equations 2.3 and 2.5:

$$\tan \beta = \frac{gR}{v^2}. \tag{2.6}$$

To counteract friction or to increase their speed, hockey players can also accelerate by skating while turning. After all, speed skaters such as five-time Olympic medal winner Bonnie Blair don't simply glide around the curves; they skate hard around them. When Blair turns, her skates are oriented at a slight angle relative to the circular path and her legs alternate at pushing sideways. The net result is both

Figure 2.7.   As Calgary Flame Valeri Bure makes a sharp circular turn toward the net, his centripetal acceleration enables him to lean inward. Leaning angles as low as 45° or less are possible. CP Picture Archive (Stephen J. Carrera).

a centripetal force and a forward force, which combine to generate a higher speed through the curve.

Braking

There are many ways to slow down—people who are just learning how to skate find that out quickly! But hockey players usually brake simply by straightening up their bodies and turning one or two skates sideways, rather like a skier makes a parallel turn. This lateral motion of the blade delivers the greatest resistive force and scrapes a thin layer of ice from the surface. (That I know from experience—some players try to annoy a goaltender by peppering his face with a spray of ice after he's made a save.) Equation 2.3, the formula for obtaining the leaning angle of a skater's body, is valid irrespective of the value of the acceleration, which is negative for a skater who is braking. The deceleration is determined by the friction force on the blade and therefore depends on, among other factors, how deep the blade penetrates the ice. A sharper blade will dig deeper and allow a shorter stop. Professional hockey players know this; they have their skates sharpened before every game and sometimes between periods.

Speed versus Experience: Who Wins?

As professional players approach the end of their careers, sometimes at the age of 40 or older, they gradually lose their skating edge and tend to rely more on their experience to stay competitive. Certainly, when Gordie Howe joined the Hartford Whalers at the age of 51, he didn't have the legs he'd had when he started some 30 years before. Even so, he was able to compete in the NHL thanks to his experience and deep understanding of the game. Younger hockey players may be quicker, but they tend to be more vulnerable and crack when under too much pressure, a danger during the NHL playoffs. Because of this, coaches generally prefer to use their more experienced goaltenders in the postseason.

In the winter of 2000, I saw firsthand how experienced players could dominate the game of hockey. A team of NHL old-timers showed up in my hometown of Moncton, Canada, to take on a local team of younger players from the local Royal Canadian Mounted Police. The NHL team consisted of such hockey legends as Guy Lafleur and Marcel Dionne, formerly of the Montreal Canadiens and the Los Angeles Kings, respectively. Most on the roster were in their 50s, and many of them were probably grandfathers already. It was interesting to see the hockey veterans playing so much better than their opponents, many of them twenty-five years younger. Even though the younger players appeared to be quicker, their speed was to no avail.

As they usually do on their road show, the former stars did not play seriously until the end, then they turned the game into a rout, much to the delight of cheering fans. We saw Guy Lafleur go on a breakaway and, instead of shooting on the helpless goalie, make a long pass backward, without looking, to his teammate Marcel Dionne, who was positioned behind the two defensemen chasing Lafleur. Dionne finished the job with a perfect wrist shot into the far corner. This kind of superior positional play, accurate passing, and shooting ability made for a convincing victory of experience and knowledge over speed. The former NHL players seemed to have, as the saying goes, "eyes on the backs of their heads."

On the other hand, there are times when speed and stamina are key ingredients to success. The Stanley Cup final of 2000, between the young and energetic New Jersey Devils and the experienced Dallas Stars, was perhaps a classic example. It was a tight, low-scoring series with an average of only four goals per game, many of the games ending in overtime. Game five was especially grueling, as it went into its third period of overtime before Dallas scored the only goal of the game. But it was a Pyrrhic victory. The game was almost twice the normal length and put a huge amount of strain on the players, who were already feeling the stress of the championship final. As exhaustion and injuries took their toll on both teams, the younger legs of the New Jersey players helped them eventually win the Stanley Cup, four games to two.

NHL teams, as a rule, try to create an ideal mixture of youthful talent and proven experience—while keeping within the payroll budget, of course.

Power

Some hockey players, like Jaromir Jagr of the Washington Capitals, are described as "power forwards," an expression referring to their ability to break through opponents. They seem to be at ease meeting resistance while maintaining good speed. To a physicist, the word *power* means something very specific. It is defined as the rate at which work is accomplished, and it is measured in units of watts (the same watt used to describe the output of a light bulb), in honor of physicist James Watt. The unit is symbolized by the letter W, and one watt is equal to one Joule of energy per second. A 100 W electric bulb thus consumes electrical energy at a rate of 100 J every second. Work, too, has a scientific interpretation (see Appendix 2 for further discussion). Work is done when a force f moves something over a distance d, and it is described by the equation $W = fd$. How much power does this force produce? Because power tells us how quickly work is done, it depends on the velocity of the mass on which it is applied:

$$P = fv. \qquad (2.7)$$

The combination of force and velocity yields power: if either f or v is null, then no power is generated.

For a hockey player, power therefore involves a combination of strength and speed. In order to move fast through resistance—and fans know how much hooking and grabbing goes on in the NHL these days—power is the key. Watching a player like Philadelphia's Mark Recchi, giving 110 percent at every shift, whether he's skating alone or struggling in a corner with an opponent, one can't help but think of how much power he is generating.

Physiologically speaking, a powerful muscle is one that is strong (that is, contains many fibers) and is able to burn energy rapidly. When a muscle contracts it exerts a pulling force. Scientists have investigated how this force depends on the speed of contraction, and Fig. 2.8 sketches a typical curve for the force produced as a function of the speed of contraction. The greatest force is delivered at slower speeds, which is typical of activities such as bench-pressing heavy weights. At higher speeds the muscle can't pull as forcefully, owing to the physiological constraints related to the way the biological energy (in the form of adenosine triphosphate, or ATP, the main energy source of living organisms) is expended. Somewhere in the middle lies the optimal speed, at which the product fv and the power output peak. For many fast-paced sports, including hockey, power, not strength, is the most important element. Special training techniques designed to develop a powerful body involve moving moderate loads as fast as possible. Contrary to popular belief, doing fewer repetitions with heavy weights is not an optimal exercise to increase muscular power. It is not uncommon to see strongly built hockey players who are slower and less effective on the ice.

So how do we build the ultimate hockey player? To get a feel for this, we need to examine what each muscle group contributes during skating. Sport scientists interested in speed skating have tackled this problem using a combination of state-of-the-art motion-analysis

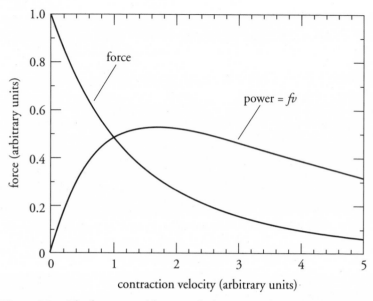

Figure 2.8. The force exerted by a muscle decreases with increasing speed of contraction. Power is greatest at some medium speed, at which the product fv is maximum.

equipment, mathematical models, and computer simulations. Their results, summarized in Fig. 2.9, show the power output at the three main joints (hip, knee, and ankle) during one push-off. We see that most of the power comes toward the end of the stride, when the joints are almost fully extended. The energy released by each muscle group is the sum of the power over time, or, equivalently, the area under each curve. Typically, the ankle, the knee, and the hip contribute to 15, 40, and 45 percent of the total energy, respectively. The ankle yields a smaller amount of energy owing, in part, to its limited range of motion.

Shall we conclude, then, that a skater's workout should put more emphasis on the hip and less on the leg and calf muscles? As far as power skating is concerned, the research seems to point in that direction, but an athlete should try to keep a good balance in the development of each muscle group. We should also keep in mind that, unlike speed skaters, hockey players use their upper body a great deal and therefore should develop it equally.

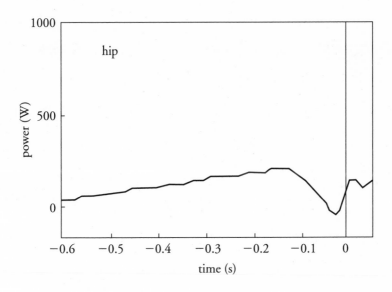

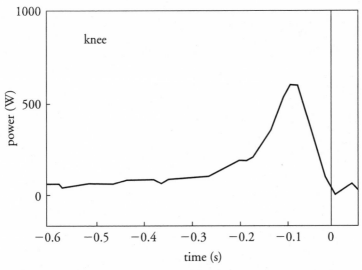

Figure 2.9. Measured power output as a function of time at each of the major joints during skating at constant velocity. The end of the push-off, toward which power is at its maximum, coincides with $t = 0$ s. From H. Houdjik, J. J. de Koning, G. de Groot, M. Bobbert, and G. J. van Ingen Schenau, "Push-Off Mechanics in Speed Skating with Conventional Skates and Klapskates," *Medicine and Science in Sports and Exercise* 32 (2000): 635.

(*continued*)

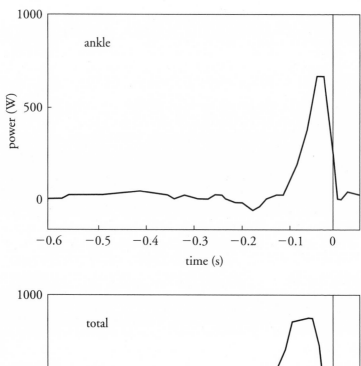

Figure 2.9. (*Continued*)

Energy Expenditure

Because hockey is such a fast game, it is a very demanding sport in terms of energy, more so than other team sports like basketball and baseball. Hockey players move faster on their feet than athletes of any other team sport; some NHL players skate at speeds in excess of 25 mph (40 km/h or 11 m/s), and this speed is more often limited by the size of the rink than a player's physical ability. The legendary Bobby Hull, the fastest skater of his time, was once clocked at 29.2 mph, or 47 km/h. After Hull had spent 29 minutes on the ice during one game, sport scientists figured out that he had skated over 13 km (and he was hardly even weary afterward). It is not surprising that during 60 minutes of regulation time a player will burn as many as 6,000 calories and lose up to 15 lbs. of body weight!

Although hockey involves a wide range of activities—puckhandling, shooting, checking, stopping, turning—moving one's own weight around is what requires the most energy. Of course, hockey players don't constantly skate as hard as they can throughout a game; rather, they typically skate at 75 percent of their top speed, waiting to see how play will evolve.

The energy spent for propulsion during skating serves to do three things: overcome the friction of the ice, combat air resistance, and increase one's kinetic energy (that is, accelerate). One of the most fundamental tenets of physics is that energy is neither created nor destroyed. This means the energy (or power) spent by the player and the energy used to increase his kinetic energy and overcome friction and drag must be equal. In mathematical terms this looks like

$$P = P_{air} + P_{ice} + \frac{\Delta K}{\Delta t}, \qquad (2.8)$$

where P is the total power generated, P_{air} and P_{ice} are the power used to overcome air and ice friction, respectively, and $\Delta K / \Delta t$ is the rate at which the kinetic energy of the skater is changing. The kinetic energy is given by $K = \frac{1}{2} mv^2$. At constant velocity, the kinetic energy doesn't change, and therefore $\Delta K / \Delta t$ is zero. Consequently, when skating at a steady speed, all of a skater's power is used to counterbalance the energy dissipated through friction.

To get any further, we need to figure out what the frictional losses are. For the ice, we know the friction force and that

$$P_{ice} = f_{ice}v = \mu mgv. \tag{2.9}$$

As seen in the previous chapter, μ, the friction coefficient of a clean surface of ice, typically ranges from 0.003 to 0.01. The friction coefficient is usually higher when the ice is rough or snow has accumulated. This has been studied in detail both by geologists interested in the formation and movement of glaciers and by traffic-accident investigators who need to know how long it takes a vehicle to stop on an icy road.

Air friction (or drag) has also been studied for many years. It has been a subject of interest over the past century because of its relevance to the aircraft industry, and more recently it has been applied to unravel the mysteries of insect flight, to explain the motion of baseballs, and to model the flow of inhaled medication inside the lungs. The drag force exerted on an object moving at a speed v in a stationary fluid is proportional to v^2, and the power needed to overcome this drag is

$$P_{air} = Cv^3, \tag{2.10}$$

where C is the drag parameter, which depends on the shape of the body, the cross-sectional area, the viscosity of the air, and other environmental parameters. C also depends on the velocity itself, which complicates things. When turbulence kicks in at high speed, for instance, C can increase significantly. However, for the range of velocities considered here, it is appropriate to consider it as constant.

Air friction at average skating speed accounts for approximately 75 percent of energy dissipation, while ice friction is responsible for the remaining 25 percent. In other words, $P_{air}/P_{ice} = Cv^3/\mu mgv = Cv^2/\mu mg \approx 3$, from which the air drag parameter C can be estimated. We can assume that for a skater of average height and size, skating at a speed of 8 m/s with $\mu = 0.005$ and $m = 90$ kg, C has a value of the order of 5.

Given these results for power dissipation due to ice and air, the velocity of the skater can be obtained at any time from the total power generated by the skater. The trick, though, is to find out what that

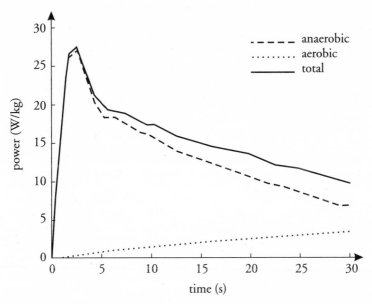

Figure 2.10. A typical power output per unit of body mass, from a group of elite athletes during a 30-second bicycle ergometer test. Aerobic and anaerobic power contributions have drastically different dynamics.

power output is. One way to do so—and it turns out to be a good measure of the overall energy spent by an athlete—is by using a bicycle ergometer. The stationary bicycle has a flywheel equipped with a device that gives the instantaneous power being produced. A graph of the power versus time is plotted in Fig. 2.10, which yields the curve corresponding to an average taken from ten elite speed skaters.[7] Surprisingly, the shape of the curves is similar for athletes of every sport, not just speed skating, and is indicative of how fit the athlete is rather than what his or her athletic specialty is. An important feature of the graph is that, only 30 seconds after maximum exertion has been reached, the power output has already dropped by half. We can now understand why NHL athletes take many short shifts on the ice (typically less than a minute each) rather than fewer long ones: this is

7. J. J. de Koning, G. de Groot, and J. van Ingen Schenau, "A Power Equation for the Sprint in Speed Skating," *Journal of Biomechanics* 25, no. 6 (1992): 573–80.

the best way to operate at optimal capacity. As the old hockey adage goes, "stay on the ice and pay the price"—there is not much point in keeping an exhausted star player on the ice if he can be outpaced by a mediocre defenseman. Exhaustion is often a problem at the un-coached amateur level, where some players annoy teammates by hog-ging the ice time even when they can no longer give their maximum.

Physiologically, the switch from anaerobic to aerobic energy pro-duction at the onset of any intense physical activity explains the sharp drop in power output. The energy already present in the muscle tissues (in the form of energy-rich phosphates) and the short-term energy production mechanisms (anaerobic glycolysis) provide a high-level but temporary energy burst. Once the reserves are depleted, oxygen-based (aerobic) energy production kicks in with the help of an increased blood flow to the muscle fibers. The aerobic process is less intense than the anaerobic one, but it can be sustained for much longer periods of time; runners rely on the aerobic process to run a marathon, for instance.

Of course, every subtlety of human motion is not taken into account in Equation 2.8. For example, it does not incorporate the energy lost during acceleration and deceleration of the upper and lower body limbs. Yet these other variations are actually not as im-portant as they may seem, since cycling also includes such loss of energy—roughly the same amount as is lost during skating. In the end, the power measured on the ergometer is close to the amount used for propulsion on ice alone.

For the advanced reader who might be interested, we can still extract some more information from Equation 2.8, though we need to use calculus, specifically to rewrite the term $\Delta K / \Delta t$ as $\Delta K / \Delta t = mva$, where a is the acceleration. Then, putting Equation 2.8, 2.9, and 2.10 together, we obtain the following:

$$P(t) = \mu mgv + Cv^2 + mva. \tag{2.11}$$

This equation can give us the location and speed of the skater at any time. We simply need to plug in the power $P(t)$ generated by the ath-lete as measured experimentally on the ergometer. With reasonable estimates for C and μ, the equation is easily solved numerically with a computer. A typical velocity curve is plotted in Fig 2.11.

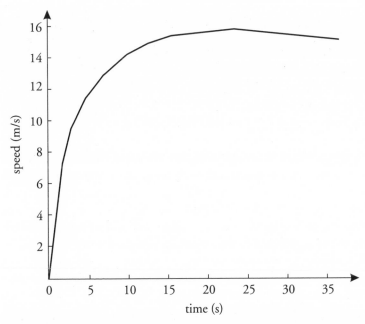

Figure 2.11. Calculated skating velocity based on measured power outputs, similar to that of Fig. 2.10 (both from J. J. de Koning, G. de Groot, and J. van Ingen Schenau, "A Power Equation for the Sprint in Speed Skating," *Journal of Biomechanics* 25, no. 6 [1992]: 573–80). This model compares well with experimental data.

The proof of the pudding, as the saying goes, is in the eating. The present model, developed by J. J. de Koning, G. de Groot, and J. van Ingen Schenau, sounds plausible, but how does it compare with reality? The answer is that the model compares quite well with actual measurements done on skaters. Equation 2.11 is more than an equation that happens to match experimental data. In it, we have a successful model of ice-skating, and with it scientists have been able to understand and predict, among other things, what factors (aerobic versus anaerobic energy production) contribute most to the performance of speed skaters on short races like the 500-meter sprint. This information enables speed skating coaches to create physical training programs that most benefit their skaters and improve their competitiveness. What's true for speed skaters, at least in this regard, holds true for hockey players as well. An understanding of the biomechanics of skating helps coaches improve the fitness level of their players and give them a better chance in the race for Lord Stanley's Cup.

Technique

Based on what's been said so far, we might infer that great strength and power are all that matter in skating. But if that were the case, how is it that, even with their smaller stature and lesser strength, the top female speed skaters easily outskate the fastest NHL player? The fact is there's more to it than strength and power: skating is also a matter of technique. Good skating technique doesn't just make for faster players, it also allows them to be more efficient as they go from point A to point B. Good skaters can stay on the ice longer and, toward the end of the game, are able to skate around wearier, slower opponents. Some of it can be learned through practice, some of it can't. The finer points of good technique involve subtle coordination and body movements that are beyond the scope of this book, but detailed discussion can be found elsewhere.[8]

In his book on hockey, Randy Gregg, former Edmonton Oilers defenseman, recounts the opinion of his wife, former Olympian speed skater Kathy Voigt.[9] After attending a game in Edmonton, she commented that the only player whose skating technique closely resembled that of speed skaters was Paul Coffey, then a member of the Oilers. Coffey was—and still is—known for his skating ability, which makes him one of the most effective defensemen in NHL history. Even though his main function on the ice is to stop opponents, Coffey's skating finesse enabled him to rack up almost 1,500 points over 20 seasons, ranking him among the most prolific players. With his speed he can venture deep into the offensive zone and still come back quickly to defend. Even late into his career, he remains one of

---

8. A detailed discussion on the technique of speed skating is found in B. Publow, *Speed on Skates* (Champaign, Ill.: Human Kinetics, 1999).

9. A good discussion of the technical and mental aspects of hockey can be found in R. Gregg, *Hockey* (Stettler, Alberta: F. P. Hendriks Publishing, Ltd., 1999).

Figure 2.12.   Boston Bruin Bill Guerin demonstrates a perfect skating technique on his way to winning the Fastest Skater event at the annual skill competition of the All Star weekend. CP Picture Archive (David Zalubowski).

the fastest skaters in the NHL, proving once more that technique is just as important as strength.

Coffey is a classic example of a player with great speed and agility, but he's not the only one. Boston Bruin Bill Guerin and Carolina Hurricane Sami Kapanen are also great skaters and among the top scorers in the NHL today.

Chapter 3

SHOOTING

All hockey players dream about scoring that perfect goal, the one that wins an important game for the team. The more spectacular the goal is, the better. Trying for a hard shot from the blue line—or, as commentators sometimes say, "going for the top shelf"—and beating the outstretched glove of a quick goalie is the kind of play that lifts a crowd.

Until now we've focused on ice and skating, but most kids learn to stickhandle and shoot before they can skate. After all, shooting or passing to teammates in a game of street hockey can be just as fun as skating around. Knowing how to shoot accurately is a valuable skill, as about three-quarters of the points are scored this way (the remaining quarter are scored by going around the goalie). One of my NHL heroes—who happens to have made the decision to return to the game after four years in retirement—is Mario Lemieux of the Pittsburgh Penguins. Among Lemieux's distinguishable skills are his stick-handling and quick shooting. One-on-one against a goalie, his magical touch is almost always successful.

Becoming a great hockey player requires mastering many aspects of shooting. For one thing, you can't hope to score unless the puck is on target, so accuracy is indispensable. Shooting speed, which is mostly a matter of technique, is also important. Physics can help us understand both of these skills, because it tells us how fast a puck will go once it is hit or pushed and where it will end up once it leaves the stick. Before going into the physics of moving pucks, let's take a look at the puck itself.

The Puck

The black dot everyone chases after in a hockey game is nothing more than a disk made of vulcanized rubber. The rule book says it must weigh 6 ounces (about 170 grams) and measure 3 inches across and 1 inch in thickness. The puck density is slightly less than 1.5 grams per cubic centimeter, a bit denser than water. True to its name, vulcanization is a process that puts rubber through hell—it is treated with heat and brimstone, nowadays known as sulfur, making the rubber hard and impervious to heat and cold. Vulcanization was accidentally discovered in 1839 by American inventor Charles Goodyear. Because natural rubber is unable to withstand temperature extremes, it is of limited use to industry. One day, Goodyear dropped a mixture of rubber and sulfur onto a hot stove; later, he found that the composite had become much harder and resistant. The recipe was refined to make automobile tires, shoe soles, and many other items—including hockey pucks.

Rubber is one of the most elastic materials on earth, and even vulcanization can't stop hockey pucks from bouncing. Smashed against a hard surface like concrete or ice, a puck rebounds with between 45 and 55 percent of its original velocity (less so on a softer surface like a board). This percentage is the so-called *coefficient of restitution*. In an ideal world, the puck wouldn't bounce off the ice at all. To minimize this unruly behavior, someone discovered a long time ago that freezing the puck before a game would make it slide better and bounce less, owing to its increased stiffness. This can be demonstrated in a simple home experiment: place a puck in a freezer for an hour and then let it drop sideways on a concrete floor, along with a puck kept at room temperature. You will find that the cold puck bounces less than half the height of the warm puck. In fact, they will bounce to about 12 percent and 27 percent of their original height, respectively. Note that a 50 percent velocity restitution versus a 25 percent height restitution for the warm puck is not a contradiction. This difference exists because the height at which the puck rises depends on its initial kinetic energy, which goes as v^2. So if we cut the puck's velocity by half, it comes back to $(1/2)^2 = 1/4$ of its original height. For the frozen puck, the coefficient of velocity restitution is therefore on the order of 35 percent, as opposed to 50 percent for the unfrozen ones. This is why buckets of pucks are kept refrigerated during NHL hockey games. Unfortunately, this creates a new problem: a hard, frozen puck traveling at 90 mph is a dangerous projectile! Even when under considerable pressure, like that applied during a slap shot, the rubber doesn't bend or squeeze appreciably.

Pucks have changed over time. Hockey was originally played with a lacrosse ball, but exasperated rink owners soon had enough of the airborne balls breaking windows. They cut the ball into three slices and kept the middle section. Ever since 1885, the game has been played with just such a rubber slice. From an aerodynamic point of view, cylindrical pucks are better than balls because they don't tend to go all over the place when hit. For the same volume, a sphere has a larger cross sectional area (that is, it looks bigger from the side), so the friction force of the air is higher.

Players and goaltenders learn to predict the puck's reaction to the ice and the stick. It is important to do so in a game where a few inches often make the difference between a goal and a save. Of course, goaltenders hate bad bounces and pucks that have unpredictable, wiggly trajectories. I would rather face a clean and fast shot than a spinning puck that bounces off the ice.

Not all pucks are identical, nor do they feel the same. This became evident during the infamous "Fox puck" saga. In an honest attempt to increase its American audience by making the puck easier to see, the Fox television network introduced a high-tech puck during the 1995–96 All Star game. The elaborate and costly technology enabled Fox to superimpose a blue, cometlike trail around the puck that turned red when it traveled beyond 70 mph. Darker shades of red appeared whenever the puck traveled faster. Inside the modified puck was a microchip that sent pulses through infrared emitters located all around the edge of the disk. The infrared signal—invisible to the eye, just like the signal that comes out of your TV remote control—was picked up by sensors placed along the rink board, which synchronized with the pulsing puck. Infrared cameras placed above the ice could then pick the signal and send the information to a "Puck Truck," a minivan loaded with computer equipment. Computers would then analyze the puck location and velocity to superimpose a halo around the puck on the TV screen.

Although the glowing puck made it easier for people to follow the game, hardcore hockey fans were not thrilled about the gizmo. Many NHL players also complained about the way it bounced off the ice. Fox insisted that all pucks had been tested and were within specification, but players claimed that the four screws on one side of the puck prevented it from sliding like a regular puck. To make things worse, the modified pucks were full of wires and sensors, making them prohibitively expensive for the players to practice with. The debate climaxed during the 1997 NHL playoff final, between Philadelphia and Detroit, when the hard-to-handle gadget exasperated the players. Amid the controversy, Fox abandoned the project, saying that at $50,000 a game, the price was just too high. Since then, all NHL games have been played with the good old rubber puck.

Now that we know what a real puck is made of, we can turn to what hockey players do with it. First we will discuss passing, which happens far more frequently than shooting.

Passing

Passing is the simplest way to move the puck, in a straight line along the ice. But fast and accurate passing is a skill that requires much practice. It's easy to forget that in addition to being a great scorer, Wayne Gretzky was also a formidable passer. This is obvious from the huge number of assists he had during his 20-year career. Without his passing ability, he would never have reached his record 215 points in a single season which he did in 1985–86 with the Edmonton Oilers. Those who were lucky enough to play with him allude to his uncanny ability to find their stick through a forest of other legs and sticks.

Typically, the puck stays in contact with the ice during a pass. According to Newton's laws of motion, if we neglect friction it should move in a straight line with an almost constant velocity. When two players are standing still or moving in the same direction at the same speed, passing the puck is straightforward; but when the players are moving relative to each other things become a bit trickier. In a breakaway, for example, a winger might pass the puck to a center who is sprinting ahead. For the play to succeed, the puck must be sent in the right direction with the right velocity. A split-second wasted by the center in retrieving an inaccurate pass might be all that a defenseman needs to catch up. Simply put, the passer has to fire the puck not where a teammate *is* but where that teammate *is going to be*. So how far ahead of the receiver should the puck be aimed? This depends on both the distance initially separating the two players and their velocities. Fortunately, basic physics can help out. If the puck is moving much faster than either player, which is often the case, then the passer aims for where the receiver is right at that moment, since the receiver won't have time to move far before the puck gets to him. The basic lesson is, if you're going to pass the puck, pass it quickly. That's just what NHL players try to do. If teammates are moving in the same direction at the same speed (in other words, if they aren't moving relative to each other, which is sometimes the case during a two- or three-player fore-check), passing is as simple as if they were all at rest. From a physics point of view, the two situations are equivalent. This is an example of relative motion, like that first examined by Italian physicist Galileo Galilei (1564–1642), who studied, among other things, free-falling objects dropped from the leaning tower of Pisa. Physics tells us that as long as two hockey players are not moving relative to each other, they simply need to aim at each other when they pass the puck; whether the ice is moving under their feet or not is not important.

Friction, as we have seen, slows things down. So the obvious question is: How much does ice friction affect passing? The stick blade is only one foot long, and passes can be as long as 80 feet, so can ice friction really mess things up? A typical ice friction coefficient is 0.005 (see Chapter 1), but on a rough and snowy surface it can be significantly higher. To make things simple, let's suppose $\mu = 0.01$. The ice friction force on the puck is then $f = \mu mg$ and produces a deceleration, according to Newton's law, at a rate of $a = f/m = \mu g = -0.1$ m/s^2 (negative because it's slowing down), meaning that with every second, the puck slows down by 0.1 m/s. As high school and college physics students know all too well, the motion of an object with constant acceleration is easily predictable. Without friction, the puck travels a distance d in time t, where $v = d/t$ is the puck speed. In other words, $d = vt$. If the puck is hit with the same speed v on a surface that had friction, then it would travel a distance $d = vt + \frac{1}{2}at^2$.

This is a well-known formula related to constant acceleration. How much farther would the first puck have traveled than the second one after a time t? The difference between the two is $\frac{1}{2}at^2$. Now t is roughly equal to d/v, so the discrepancy is $\frac{\mu g}{2}\left(\frac{d}{v}\right)^2$.

Let's take an example. A 60-foot-long pass (roughly from one blue line to the other) usually takes about a second, which corresponds to a puck velocity of 20 meters per second. At that same speed, an 80-foot pass—one of the longest you can legally make—would have to be corrected by about 8 centimeters (3 inches). Such a small error certainly wouldn't bother an experienced stickhandler like Anaheim's Teemu Selanne. This means that ice friction doesn't really affect passing; on the other hand, you won't be able to blame the ice for bad passing (unless it's bumpy). To sum it all up, ice friction is an issue only at very low passing velocities or on badly damaged ice, neither of which is usually a factor in the NHL.

In a real three-dimensional world, however, a puck doesn't always slide smoothly. It can also bounce, roll, spin, and wiggle all kinds of ways. We talked earlier about a bouncing coefficient of about 12 percent for a frozen puck. Theoretically, the puck should then stop bouncing after two or three hits, for after the third bounce it would rise to only 0.2 percent of its original height. Unfortunately, it doesn't bounce as predictably as, say, a rubber ball. The asymmetrical shape of the puck greatly affects how it rebounds. Much like when a football hits the ground, there is just no way to predict how the puck will behave from one bounce to the other. In physics, this is called a chaotic system, something that is only predictable for a short period of time. A minute change in the initial conditions, like the puck's tilt or velocity, can dramatically change the outcome. Although chaos brings physical prediction to its knees, it sure makes life and hockey exciting!

The Mysterious Spinning Puck

Here's an oddity: when a puck is sliding straight along the ice (pure linear motion), the time it takes to stop increases with its initial velocity. When it is given a spin only (pure rotational motion), the stopping time also increases with its initial spinning velocity. Nothing surprising here. However, when we combine rotational and linear motions, things become strange. At first, one would think that each type of motion would behave independently—that is, rotation would not influence translation, and vice-versa. As it turns out, the time it takes for the puck to stop spinning *and* sliding is always exactly the same! In other words, no matter how fast you spin the puck, and how fast you send it sliding on the ice (or on any other smooth surface for that matter), it will always stop spinning and moving all at once. This seemingly weird behavior was investigated in detail in a paper by K. Voyenli and E. Eriksen.[1] Although it can easily be observed in a home experiment, a theoretical explanation is not as obvious. It requires solving quite a few ugly equations that are beyond the level of this book.

However, there's a prettier—and deeper—explanation. We simply need to invoke the uniqueness property of the laws of mechanics: any given set of conditions of a physical system (velocity, position, etc.) is usually sufficient to predict its future and know its past. This statement isn't surprising. After all, if there were many possible outcomes or histories for a given set of circumstances, physics would be totally useless in making predictions. This uniqueness property dictates the way the puck must behave. For example, if the puck were to stop moving along the ice before it stopped spinning, there would be no way to know at that point whether it initially had a linear motion or not. One observer could conclude that it never had a linear motion, while another could claim that it was moving linearly but had stopped some time ago. With this in mind, it makes sense that the puck should come to a complete rest all at once. This is due to the strong coupling that exists between the rotation and linear motion of a puck.

When the Puck "Goes Ballistic"

A projectile shot into the air is nothing more than a mass under the sole influence of gravity and air friction. Figuring out the path it takes

---

1. K. Voyenli and E. Eriksen, "On the Motion of an Ice Hockey Puck," *American Journal of Physics* 53, no. 12 (1985): 1149.

is an exercise in what physicists call *ballistics,* the science that studies the trajectory of free-falling bodies. It applies to analyzing rockets, bullets, and other warfare projectiles, and has been a field of study for many centuries. As a matter of fact, books on the range of mortars were among the first to be published after William Caxton set up the first printing press in England.

When a puck is shot off the ice, the questions that follow are: Where will it end up? Will it reach the target or not? The easiest way to analyze projectiles is to neglect the air resistance. Although hockey is not played in a vacuum, we can gain much insight by first assuming there is only one force acting on a puck—gravity.

The motion of a projectile in three-dimensional space is sometimes difficult for people to understand. Projectiles move vertically as well as horizontally; nonphysicists (and many physics students) often confuse the two types of motion or don't grasp the relationship between them. Perhaps a simpler way to approach the problem is to consider the "projectile drop," or, in our case, the "puck drop."

The first thing to realize is that the vertical and horizontal motions of a projectile are independent of each other. Gravity is directed downward, so it only influences the vertical location (or the height, represented as y), not the horizontal (represented as x). This means we can write two separate equations for x and y. The vertical acceleration is due to gravity, g, and is constant. It is not surprising, then, that we obtain the same type of equation as the one describing a puck slowing down on ice:

$$y = v_y t - \frac{1}{2} g t^2, \tag{3.1}$$

where $g = 9.8$ m/s^2 and v_y is the initial vertical velocity of the puck, which depends on the angle of the shot. For x we have no horizontal acceleration, hence the simpler equation:

$$x = v_x t, \tag{3.2}$$

where v_x is the horizontal velocity, which also depends on the angle of shooting.

The last piece of Equation 3.1 is the interesting thing here. Without it (say, if gravity were null), y and x would just increase steadily over time and the puck trajectory would be a simple straight line. But because of gravity, the puck drops from that straight line by an amount going as t^2. This is of utmost importance for hockey sharpshooters, because it means that a shot aimed straight at a target will always hit under it.

How much will the puck drop below target? Before figuring this one out, it would be nice to eliminate the time variable in the equations—after all, we don't generally carry a stopwatch with us to play hockey. Instead of using time, it would be simpler to use the distance traveled, x, the shooting velocity, v, and the shooting angle θ. Using elements of high school trigonometry, we find $v_x = v \cos \theta$, with v being the initial shooting velocity. Putting this into Equation 3.1 we get $t = x/v \cos \theta$. The puck drop then becomes

$$\frac{gx^2}{2v^2 \cos^2 \theta}. \tag{3.3}$$

Fig. 3.1 illustrates the trajectory of a free-falling puck. To take a realistic example, let's suppose the following scenario: Rangers defenseman Brian Leetch stands at the blue line and sees a nice opening at the top corner of the net. The net is 60 feet away and stands 4 feet tall, so in a flash he gets out his pocket calculator and figures that the shooting angle should be $\theta = \arctan (4/60) = 3.8°$. He then snaps a good shot

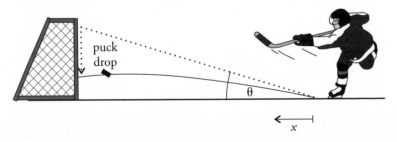

Figure 3.1. If air friction is neglected, a puck shot into the air follows a parabolic trajectory due to the effect of gravity. The curved path can be understood in terms of a vertical puck drop (dotted arrow) from the line of target.

at 90 miles per hour in that direction. According to Equation 3.3, once it has arrived at its destination, the puck will be 40 inches below where it was aimed at! Sound like a lot? It is, but in real life, when players shoot they don't think about equations and physics, they simply know from experience that the puck will drop and compensate by aiming higher.

Air Drag

Without air, mechanics is simple enough. But how do we account for the annoying reality of air resistance? Some readers may already have an idea of how air affects the motion of most projectiles. When Tiger Woods hits his golf ball, it first appears to follow a smooth parabolic trajectory, but, as it goes on, the ball slows down and falls more vertically. As a result, when it reaches the fairway it doesn't bounce and roll as much as one might expect. This kind of behavior is typical of many projectiles that are influenced by air, including a puck.

In sports that feature quick-moving projectiles, air drag sometimes becomes an issue. It significantly affects the trajectory of a baseball. At Coors Field, home of the Colorado Rockies, the air is thinner—as it is everywhere in Denver, the Mile-High City. This lower air density reduces the drag, and, consequently, more home runs are hit at Coors Field than at other stadiums of the same size. This doesn't have much to do with the Rockies' starting lineup—they hit fewer home runs when on the road, and visiting teams hit more runs at Coors Field than anywhere else.

The International Table Tennis Federation recently decided to increase the official ball size from 38 to 40 mm. The hollow plastic ball is so light that this modest increase in cross-sectional area lowers the ball speed by about 10 percent, making the game easier for spectators and TV viewers to follow. (Wisely, no one proposed a "Fox Ping-Pong ball" solution!) The International Tennis Federation is now toying with a similar idea. Slowing down tennis balls might eliminate the dominance of power-serve players and bring more volleys, net approaches, and extended rallies back into the game.

Is air drag also an important element in hockey? To find out, let's consider the simplest case of a shot fired along the ice.

Puck Drag on a Flat Stretch

The drag force, labeled f_{drag}, depends on the velocity of the projectile as well as its area and the properties of the surrounding fluid. More precisely,

$$f_{drag} = \frac{1}{2} C_D \rho A v^2, \tag{3.4}$$

where ρ is the density of the fluid (1.22 kg/m^3 for room-temperature air at sea level), A is the cross-sectional area of the projectile, and C_D is the drag coefficient, which depends on the shape, texture, and velocity of the projectile. Drag coefficients have been measured experimentally in wind tunnels for a wide variety of shapes and can be found in standard textbooks on fluid mechanics.[2] Drag coefficient is an important parameter in the design of aircraft. As Fig. 3.2 illustrates, there are two drag coefficients to be considered on a puck: one for the side, from which the puck appears as a cylinder, and one for the top, from which it appears as a flat disk. The cross sections for each case are $A_1 = 19$ cm^2 and $A_2 = 46$ cm^2.

Unfortunately, drag coefficients for a cylinder with a diameter-to-length ratio similar to the puck are not usually listed in fluid mechanics references. This means we have to come up with an answer ourselves (which is part of what makes physics fun). But attempting to theoretically calculate the drag on something like a puck is a tedious enterprise, and, because of the assumptions one needs to make in order to simplify, the end result is not always accurate. It is much better to go into the laboratory and actually measure it, which is what I did. By putting a puck on a cart atop an air-track—an apparatus that creates an air cushion to produce very little resistance—and subjecting it to a constant wind, we can calculate the drag force based on the puck's acceleration. Drag coefficients of $C_1 = 0.46$ for the side and $C_2 = 0.56$ for the top were obtained, with an experimental error of the order of 10 percent. The hockey puck is therefore less aerodynamic (and streamlined) than a Honda Prelude ($C_D = 0.39$), but more so than a thin circular disk facing

2. W. F. Hughes and J. A. Brighton, *Fluid Dynamics* (New York: McGraw-Hill, 1999), 116.

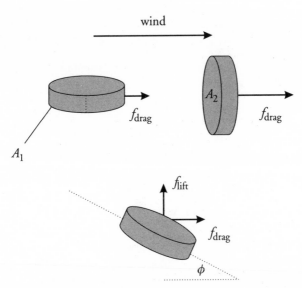

Figure 3.2. Air drag affects the puck's motion in two ways. First, the drag force tends to slow it down horizontally, and, second, when a puck is tilted at positive angle ϕ relative to the ice, there is a lift force directed upward. Each force depends on the drag and lift coefficient as well as the wind velocity.

the wind ($C_D = 1.16$). It stands in the same category as a solid hemisphere with the curved side facing the wind ($C = 0.42$).[3] (For those who might be interested in experimental details, the wind speed for my measurements was 4.6 m/s, giving a Reynolds number of about $R \sim 10^4$. Because drag coefficients tend to remain constant for $10^3 < R < 10^5$, my results are applicable for the top puck speeds encountered in hockey. A few measurements were made at 10 m/s in our wind tunnel facility, which also yielded drag coefficients near the 0.5 mark, so I am confident in my results.)

   With this information in hand, we now put the drag force into the equation of motion and calculate how fast the puck will go at any given time. To do so, we need to solve a differential equation, a procedure that is beyond the scope of this book. Rather than go the whole nine yards, a simpler way to appreciate the air drag is to calculate the drop of velocity per unit of distance traveled. This can be done using the principle of energy conservation, which says that the work done on the puck by the drag force is equal to the puck's change in kinetic energy. We'll bypass the details and get to the answer right away:

$$\frac{\Delta v}{\Delta x} \approx -\frac{C_1 \rho A_1}{m} v = -6.3 \times 10^{-3} v. \qquad (3.5)$$

This equation gives the velocity change of the puck per unit of distance with v expressed in m/s. Notice that the deceleration is more important at high velocities, when air resistance is greatest. In the fastest shots in NHL games, pucks move at a speed of about 100 mph; in such cases the puck slows down by 0.3 m/s every meter, or 0.2 mph every foot. After traveling from the blue line to the net, 60 feet away, the puck would be approximately 12 mph slower, an appreciable amount. In comparison, if the puck had started out at 60 mph the velocity drop would be 7 mph over the same distance. In conclusion, air drag really does influence the puck velocity on a long range, and it is more pronounced at high speeds.

Can a Puck Fly?

We now turn to the problem of taking air friction into account when the puck is shot into the air. Pucks shot high are common these days, with so many goalies using the butterfly technique popularized by Patrick Roy in the 1980s. The technique exposes very little space at the bottom of the net—the traditional weak point of a goaltender—so a shooter is often better off going for the sides and the top corners of the goal. Experts say that even when a goalie is standing up, the ideal shot would still be a couple of inches off the ice, just above the lower part of his stick.

What happens, physically speaking, when the puck is lifted off the ice? How does air resistance come into play? Once again, we need to split the motion into two parts. The first part is the horizontal motion, which takes place parallel to the ice, and the second is the vertical motion. The horizontal speed is much greater than the vertical speed because the upward angle at which the puck is hit is usually quite small. For instance, if a player aims at the top part of the net from the blue line, the vertical velocity is 7 percent of the horizontal velocity. Therefore, air drag affecting the up-and-down motion is not likely to be a factor in determining the vertical location of the puck. On the other hand, if the puck is tilted (that is, not parallel to the ice), things change. As with any asymmetrical object, we need to consider the lift force acting on the puck. If the puck moves sufficiently fast, can it be lifted like an airfoil? As discussed earlier, we could try calculating the lift force using aerodynamic theory. There are simple formulas giving the lift coefficient of airfoils, but the problem is that the shape of a tilted puck is closer to a Rubik's Cube than the wing of an airplane. Because the puck has a height-to-length ratio of only 1:3, it does not act like an efficient wing. The wind pressure on the front surface of the puck can't be neglected. In principle, a small positive tilt (known as the "angle of attack" in aerodynamics books) could result in a net force on a puck that is directed downward, like some kind of suction. Such "negative lifts" are exploited in auto racing to keep cars from becoming airborne. (Some high-speed Formula One cars, going fast enough, could actually be driven upside-down on a ceiling!) Again, the best way to determine the lift force on a puck is to simply measure it.

Because the lift, just like the drag, scales with the velocity squared, we may write the following:

$$f_{lift} = f(\phi)v^2, \tag{3.6}$$

where $f(\phi)$ is the lift factor (not the same as the lift coefficient), which takes into account the cross-sectional area of the tilted puck. Experimental results (plotted in Fig. 3.3) were obtained with a similar apparatus as the one described earlier. They show a lift increasing with the tilt angle but peaking at around 25°. Measurements were done with a wind speed of about 10 mph. For convenience, a simple fit to data was done (shown on the graph), giving:

$$f(\phi) = 4.0 \times 10^{-5}(\phi) - 7.8 \times 10^{-7}(\phi)^2, \tag{3.7}$$

where f is in units of newtons and (ϕ) is expressed in degrees. Because of experimental error, Equation 3.7 is only valid to within 20 percent. For greater accuracy, measurements should be carried out in a wind tunnel at higher wind speeds, but, for the sake of this discussion, our

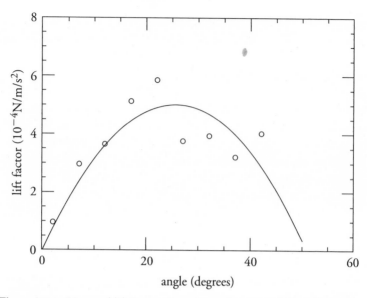

Figure 3.3. Measured lift factor of a puck as a function of tilt angle. Wind velocity is 4.2 m/s (9.5 mph).

rough estimate of the lift force is sufficient. My data are supported by a study on the aerodynamics of discuses used at track and field competitions, in which the author also finds a lift coefficient peaking near 25°.[4] The main difference is that the discus is flatter and more rounded at the edges than a puck (it has a diameter-to-thickness ratio close to 5:1), resulting in a lift coefficient that is up to three times larger. This large lift force actually allows athletes to hurl the discus farther against a moderate wind than against a tail wind, a phenomenon that has been observed both theoretically and on the field. Although it remains relatively small and constant for angles of attack up to 20°, the drag coefficient was also found to influence the discus flight.

Now we know enough to deal with the question of whether a puck can fly. The short answer is that it's not likely. If the puck is hit along the ice while tilted at its optimal angle (a bit like the wing position

4. C. Frohlich, "Aerodynamics Effects on Discus Flight," *American Journal of Physics* 49, no. 12 (1981): 1125.

when an airplane takes off), the lift force becomes greater than gravity at a speed of about 118 mph. Some of the hardest slap shots by St. Louis's Al MacInnis are still about 15 mph bellow this speed. In order to reach that kind of speed the player must hit the puck cleanly. As any player will tell you, shooting a puck at 100 mph is hard enough, let alone giving it a nice 25° angle in the process! Even if the puck were hit that fast at that angle, the drag force would quickly bring it down to sub-flight velocities (decelerating by 10 mph within 50 feet, according to an estimate based on the above measurements), so it wouldn't be airborne for very long. So although a flying puck is a theoretical possibility, it is not likely to happen.

Air Drag and Puck Drop

Even though flying pucks are impractical, aerodynamics will still influence their trajectory. Determining the exact puck drop with air resistance factored in is tricky because the lift force is sensitive to the wind angle relative to the puck, which happens to change as the puck goes up and down. And if the puck wobbles, then the effect of air drag and lift is anyone's guess. We can nonetheless estimate the effect of air friction based on the assumption that the puck is hit cleanly. We first suppose that the (horizontal) drag and the (vertical) lift forces act independently. In other words, we assume the presence of two competing effects: first, that the puck slows horizontally because of drag, thus increasing the time it takes both to reach its target and to drop vertically, and, second, the lift force that tends to keep the puck in the air, reducing the drop.

Let's use an example to appreciate the contribution of each effect. A puck is shot at 80 mph from 50 feet away with a tilt angle of 5°. The fall without any air resistance would be 35 inches, according to Equation 3.3. Taking into account drag only, we find that the final speed would be 72 mph. Averaging this drag over the whole trip, we find that the puck would take a bit longer and drop further by about 4 inches. Now including the lift force reduces the downward acceleration so that g effectively becomes $g - f_{lift}/m$, we find that the drop would then lessen by 6 inches due to the upward force. Putting drag and lift together, we can determine that the puck will hit 32 inches instead of 35 inches below target. In this particular case, the overall effect of air drag is to make the puck slightly more airborne. In general, we find that moderate tilt angles reduce the puck drop over short distances.

Having studied the physics of a moving puck, we can now turn to the techniques hockey players use. There are two basic techniques for shooting a puck: you can sweep it with your stick or strike it hard. We will examine how the two styles of shooting differ technically and physically, starting with the slap shot.

The Slap Shot

Invented by Bun Cook of the New York Rangers in the 1930s and perfected by Bobby Hull in the 1960s, the slap shot is a violent collision between the stick and the puck. It is used whenever sheer speed is the objective. A player has a better chance of beating a goalie from the blue line with a fast slap shot than with a wrist shot. The extra speed, however, comes at a price; a slap shot is less accurate, so it should not be used in all situations.

The slap shot is certainly the most popular with fans, because it is spectacular to watch. At the end of game three in the 1997 playoff series between St. Louis and Detroit, Blues defenseman Al MacInnis fired a slap shot from center-ice past goaltender Chris Osgood to tie the game. As four-time winner of the NHL hardest shot competition, MacInnis proved once again that the slap shot is the best tool for scoring from afar.

The mechanics of a slap shot involve three stages. First, the upper body winds up and begins an accelerated rotation of the torso and the stick. Next, the stick blade makes contact with the ice and the puck, causing the stick shaft to bend and accumulate potential energy (much like a loaded spring). In the last stage, the puck accelerates and leaves the blade as the stick returns to its original shape (if it doesn't break, that is). The biomechanics involved in the motion is quite complex, but, as always, using physics we can devise a simple model to help us understand the process.

We start by assuming that the upper body and the stick rotate together with the same angular velocity about a certain point—the pivot, or fulcrum—located somewhere near the player's center of

Figure 3.4. Al MacInnis demonstrates his slap shot during the hardest shot competition at the 2000 All Star weekend. He won the contest with a shot clocked at 100.1 mph. Notice the curvature of the stick shaft as he follows through on his powerful swing. CP Picture Archive (Kevin Frayer).

gravity. The stick's plane of rotation is not as vertical as a golfer's iron and not as horizontal as a baseball swing, but it lies somewhere halfway. The orientation of this plane is not important as far as the following physics is concerned.

Angular velocity is measured in radians (or degrees) per second and is indicative of how fast an object spins on itself. In days gone by, vinyl records spun at either 78 rpm (revolutions per minute) or 33 1/3 rpm. Nowadays, instead of rpm, scientists use units called radians per second. A full cycle equals 360°, or 6.3 radians, thus records spin at 8 or 3.5 rad/s. The innermost part of the record accomplishes the same number of rotations as the rim, so the angular velocity is the same everywhere. This is not the case, however, for the absolute (linear) velocity: the further away from the center a point is, the faster that point will move. So if player and stick rotate together, the end of the stick has roughly double the linear speed as the player's lower hand. That's because the stick blade is about twice as far away from the pivot as the lower hand. All objects moving in a straight line have *linear momentum,* whereas rotating objects possess *angular momentum.* The idea in a slap shot is to convert the large amount of angular momentum carried by the player and the stick (and some linear momentum if the player is skating forward) into linear momentum for the puck.

Collision problems in physics are solved by making use of the principle of conservation of momentum. The plan is to work out what the momentum is *before* the collision, calculate what it is *after* the collision, and suppose the two are equal. In the present case, since we are dealing with rotating objects, it's the total *angular* momentum that remains constant. Angular momentum, labeled L, is defined as $L = I\omega$, where I is a quantity called the *moment of inertia* and ω is the angular velocity. This equation alone, however, doesn't get us very far, as we don't know what the moment of inertia is.

Linear momentum is $p = mv$, the mass m indicating how much inertia or resistance to motion there is. By analogy, the moment of inertia depends on the mass distribution of an object and gives an idea of how hard that object is to spin. The further away from the pivot the mass distribution is, the larger the moment of inertia. Most physics textbooks use as the example a figure skater spinning with her arms outstretched. She has a fairly high moment of inertia because her arms are a long way from her center of rotation. When she brings her hands in toward herself, she has a much lower moment of inertia and therefore spins far more quickly. This is a consequence of the principle of conservation of angular momentum. Mathematically, if $L = I\omega$ is to remain the same when I is made smaller, then ω has to increase.

The moment of inertia of a player making a slap shot depends on body mass, positioning around the stick, the mass and length of

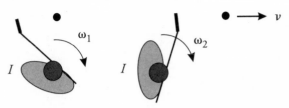

Figure 3.5. During a slap shot, the upper body and the stick rotate and collide elastically with the puck. We can create a simple model using a rotating player with moment of inertia I colliding with the puck at rest; the total angular momentum must be conserved.

the stick, and so on. As it is quite complicated to calculate, we shall simply call it I for now. Fig. 3.5 illustrates our simplified model. We also need to consider the moment of inertia of the puck. This value is simpler. Because the puck is small compared to a hockey player (it weighs 6 ounces, whereas Georges "the Train" Laraque, an "enforcer" for the Oilers, weighs in at about 240 pounds), we will treat it like a small black dot of mass m at one distance r from the pivot. In such a case, the moment of inertia is mr^2. With these approximations, we are set to write our equation for conservation of angular momentum during the slap shot:

$$I\omega_1 = I\omega_2 + mr^2\omega_3, \qquad (3.8)$$

where ω_1 and ω_2 are angular velocities of the player before and after the slap shot and ω_3 is the final angular velocity of the puck. Even though the puck does not take a circular path after the collision, its linear velocity v_3 is related to ω_3 via the equation $v_3 = r\omega_3$ (with ω_3 expressed in rad/s). In writing Equation 3.8, we assumed I (the moment of inertia of the player) remains constant during the collision, which is justified in view of the brief moment of time during which the collision happens.

But there are still too many unknowns in Equation 3.8, ω_2 in particular. If there is no permanent deformation after the slap shot (if the stick doesn't break, that is), the total kinetic energy is also conserved. The kinetic energy of a rotating body is $K = \frac{1}{2}I\omega^2$ (can you see its similarity to linear kinetic energy?), and for the puck it is simply $\frac{1}{2}mv^2$. From this a second equation is obtained:

$$\frac{1}{2}I\omega_1^2 = \frac{1}{2}I\omega_2^2 + \frac{1}{2}mr^2\omega_3^2. \qquad (3.9)$$

We now put together Equations 3.8 and 3.9 to find the final velocity of the puck:

$$\omega_3 = \frac{2I}{I + mr^2}\omega_1. \qquad (3.10)$$

We can learn a number of important things from this formula. Most obvious is the fact that the puck speed is proportional to the spinning velocity of the player. That makes intuitive sense. Second, the larger I is, the faster the shot, and this is where technique comes into play. As one might expect, swinging the stick like a baseball bat won't put very much body momentum behind it. Taking a posture that puts a lot of body weight on the stick and holding the stick lower (further away from the player's center of mass) is much more effective. This is evident in Fig. 3.6, where we see former Sabre Michael Peca putting almost his entire body weight behind the stick.

The important "body weight transfer" hockey players refer to is created by moving the body in the direction of the shot. When an NHL player winds up, most of his weight is on his back leg (the one further from the target). As the stick swings by, his body weight gradually shifts toward the front leg until he is fully supported on that leg. At that point he raises his back leg to steady his upper body and help keep his balance, rather like a pitcher raising his back leg after he finishes throwing the ball. This back-to-front motion, along with the rotational movement, reinforces the player's momentum and gives the puck its kinetic energy. If the player is also skating toward the puck, part of that linear momentum will be transferred to the puck as well. According to expert shooter Al MacInnis, simply taking one or two strides can increase the puck speed by 20 mph.

There is, however, something odd about Equation 3.10. Because a player is much heavier than a puck, we expect $I \gg mr^2$, allowing us to reduce the equation to $\omega_3 \approx 2\omega_1$. This would impose a serious limitation to the puck velocity. But in fact this assumption is a red herring. We are not supposed to compare the masses but the moments of inertia, which are a combination of mass *and* distribution. Because the puck is a long way from the pivot, whereas the player's mass is

Figure 3.6. Michael Peca convincingly shows the importance of stick flexibility during a slap shot. At the peak of the impact, the stick can bend by as much as 30°. CP Picture Archive (David Duprey).

much closer to it, his moment of inertia is not that much larger. We can roughly guess how the two compare. Suppose half the mass of a 90-kg player (basically, his upper body) is rotating during a slap shot. Let's also assume that the player is shaped like a cylinder with a radius of 20 cm (this is quite a simplification, but from a physics point of view, it's okay). The moment of inertia for a uniform cylinder of mass M and radius R is $\frac{1}{2}MR^2$. So the upper part of our cylindrical player has a moment of inertia of about 0.8 kg · m². How about the puck? It has a mass of 170 g and is located roughly 1 m from the player's pivot, so its moment of inertia is 0.17 kg · m². According to Equation 3.10, we then have $\omega_3 \approx 1.6\omega_1$. Consequently, since the ratio between the angular velocities is not quite 2, the technique (how the player postures himself around the stick) does play a role. If I increases, the puck will go significantly faster. In the best-case scenario, a very heavy player with a perfect technique would approach a ratio of 2, and the puck velocity will be limited only by swinging speed.

Our slap shot model helps us come up with sensible answers. Sometimes, unfortunately, by using models we overlook some interesting details. For example, in still photos of slap shots we can see a noticeable bending of the stick (see Figs. 3.4 and 3.6). This suggests that the force of contact between the blade and the puck is quite large, but our model doesn't say anything about that force. We can estimate the impact force by assuming the puck is under the influence of a constant force applied over a contact distance d. The ideal distance, hockey experts claim, should be about one foot. The force of impact F produces an amount of work equal to Fd on the puck, which is converted into kinetic energy so that $Fd = mv^2/2$, or $F = mv^2/2d$. This formula gives a force of 560 N (125 lbs.) for a puck hit at 100 mph and 200 N (45 lbs.) for one at 60 mph. The force may vary widely within the duration of contact, but at least we get an idea of the magnitude of the force involved.

Another important consideration overlooked is the interaction of the stick and the ice. According to our model, no matter what the player's technique or weight, if a puck is to travel at 100 mph, the blade must move at least 50 mph and the arm (at the halfway point) must swing at 25 mph. While not every player can achieve such speed, this is within the norm for pro leagues. But regardless of speed, it is a good idea to allow the blade to hit the ice just before the puck.

When it hits the ice, the stick bends even more and becomes loaded with energy that would otherwise be wasted. (Ideally, one should hit the ice one-half to one foot behind the puck for optimal results.) When the blade touches the puck, the stick shaft springs back and releases its energy to the puck. Described in terms of the conservation of energy, this energy is taken away from the player's rotational motion and transferred to the puck. The stick in this scenario may be idealized as a spring between the player and the puck. If the athlete loads the spring by hitting a wall (the ice) before the puck, then the impulse given back by the spring may be such that the player comes to a complete rest. In theory, all of the player's energy could be transferred, putting a higher upper limit on the puck velocity:

$$v_3 \leq \left(\frac{I}{m}\right)^{1/2} \omega_1. \tag{3.11}$$

In our previous example, this relation reduces to $v_3 = 2.2\omega_1$.

The contact distance in a slap shot can be used to estimate the duration of contact. Assuming the puck accelerates uniformly, we have $t = 2d/v$. This gives a contact time of 0.013 s for a puck shot at 100 mph and 0.022 s at 60 mph. Clearly, with such a short contact time, good puck control is difficult. Either you hit it right or you don't, there's no to time to change course. The laws of physics suggest—and experience confirms—that slap shots are not the most accurate way of shooting and are not recommended at short distances, where a more accurate and quickly released wrist shot is more effective.

Even though physics helps us understand how things work, firing a good slap shot takes practice above all else. I have seen countless lightweight players shoot much faster than stronger ones, simply because they had the proper technique. There's a lot of fine-tuning to do before a player can claim to master the slap shot. For example, the puck must be hit with the center of the curved blade, otherwise the stick flips and cannot generate the same amount of force. Some points like that are obvious, but it's nice to see that physics often agrees with intuition.

The Bobby Hull Enigma

In physics we sometimes come across what is called a "bad data point." This is a measurement that doesn't quite fit in with the others or line up nicely on a graph with the rest of the points. Similarly, in the world of hard-shooters, there are the modern athletes such as Al Iafrate and Al MacInnis, who each have won the hardest shot competition at the annual All-Star skill competition. Their frightening slap shots have been measured in the 100–105 mph range. These are incredible speeds, considering that most NHL fast-shooters only

Figure 3.7. Powerful defenseman and former Maple Leaf Danny Markov demonstrates the benefits of grabbing the stick low on a slap shot. CP Picture Archive (Peter Power).

reach the low 90s and the majority of NHL players never actually shoot above 90 mph.

Then there is Bobby Hull, the hockey legend from the 1960s whose slap shot had been clocked at 120 mph! He comes across as an oddity, indeed. Famed Toronto goaltender Johnny "the China Wall" Bower was quoted saying: "Stopping one of Hull's shots with your pads is like being slugged with a sledgehammer." Longtime Montreal goalie Jacques Plante was in equal awe: "His shot is like a piece of lead. One of his hard shots would break my mask if it hit it. I've caught one on my arm and it was paralyzed for five minutes afterward. Sometimes it drops five or six inches. You have to see it to believe it." Some goalies would try to grab one of Hull's bullets with their glove, only to see their wrist snap backward and watch the puck escape into the net.[5]

How could Bobby Hull have shot so much faster than current NHL players, considering that equipment, training, and fitness and nutrition programs have greatly improved since the 1960s? Was it a matter of technique, strength, or speed? One explanation proffered is that Hull's sticks were heavier than those permitted today. Made with a fiberglass shaft, they weighed 20 oz. (750 g), far more than the legal limit of 13 oz. currently in effect in the NHL. Could this be enough to give the puck an extra 20 mph? We can check this out with physics. Using a heavier stick increases the total moment of inertia I of the player. A uniform stick of length L and mass M has a moment of inertia of $(ML^2)/3$. Because the end of a stick is roughly 1 m away from the pivot, the contribution of the 13 oz. stick is 0.17 kg m^2, compared with 0.26 kg m^2 for Hull's stick. If we rely on a ballpark estimate of the player's upper body moment of inertia, say, 0.8 kg m^2, then using a heavier stick would raise his total moment of inertia from 0.97 to 1.06 kg m^2. Plugging this in to Equation 3.10, we find that the overall difference in puck velocity is only 1 percent. So, according to the laws of mechanics, we can safely rule out the idea that Hull's secret was simply his heavier stick.

The real reason must lie elsewhere, as his opponents could probably tell you. For one thing, Bobby Hull was a formidable athlete with a natural build for hockey. His wrist shot had been timed at 105 mph,

5. J. Hunt, *Bobby Hull* (Toronto: McGraw-Hill Ryerson, Ltd., 1966).

and even his backhander reached 96 mph, ten miles faster than the forehand of the average NHL player of his days. The strength of his massive torso, arms, and legs was reflected not only in the speed of his shots but also in his skating power, which made him the fastest skater of his time (at 29.3 mph). His wrists measured 9 inches around and his biceps measured 15.5 inches, both larger than those of boxer Muhammad Ali. But perhaps none of this would have mattered without a good work ethic and a lot of practice. Hull went through countless buckets of pucks, working hard at perfecting his technique while growing up in the small town of Point Anne, Ontario.

Maybe we won't see another hockey player like Bobby Hull, maybe we will. But until then, he will remain in a class of his own, a great player but a bad data point!

The Wrist Shot, the Slap-Wrist Shot, and the Backhand Shot

A wrist shot is a style of shooting that capitalizes on precision and the element of surprise. When John LeClair of the Philadelphia Flyers rushes past the blue line and winds up for a slap shot, a good goaltender has enough warning to prepare himself for a fast incoming puck. Given enough time and an unobstructed view, NHL goalies will make the save 95 percent of the time. And inside the "slot" (30 feet of the net), wasting time winding up for a slap shot might allow an opponent to reach in with his stick and deflect the puck away. To a goalie, a quick wrist shot close to the net is harder to predict and therefore more likely to succeed. The wrist shot was a favorite of the late Maurice "Rocket" Richard, the first player to ever score 50 goals in one season. "The key to scoring," he once said, "is to shoot when the goalie expects it the least." It's a simple philosophy, but it worked.

Wrist shots are not as fast as slap shots because the energy is released using a sweeping motion rather than using a long winding-up motion. The amplitude of motion is about half or less than half that of a slap shot, so the body can't build as much speed and energy during the sweep. However, because the puck is in contact with the blade for a longer period of time, it is easily guided in the proper direction with a flick of the wrist.

Some hockey experts make a distinction between the longer sweep shot and the usual wrist shot. In a sweep shot the puck touches the blade for over four or five feet, but the wrist shot is performed in as little as one or two feet. Which technique a player uses depends on how much time is available and how fast the shot should be. Of course, the shorter the sweep, the slower the puck. The speed of a wrist shot can be estimated from the distance over which the force is applied. We simply use our standard formula $Fd = \frac{1}{2}mv^2$ and assume the force is constant over the sweep. For a 5-foot-long sweep, the force needed to accelerate the puck to 105 mph—like Bobby Hull did—is the equivalent of only 30 lbs. This doesn't seem like much, but try doing it! The problem is, the puck is light and offers little inertia, so its acceleration must be very large. Pushing 30 lbs. (or 60 lbs. at the halfway point on the shaft) requires a lot of power. The kinetic energy of a puck at 105 mph is 185 J, and the time it takes to accelerate the puck over the space of 5 feet is 0.06 s. The average power delivered by Hull was therefore 185 J in 0.07 s, which equals 2,700 W, or almost 4 hp! As any sport scientist will tell you, this is a very respectable amount of muscle power, close to the physiological limits of a human being. At peak, an athlete will deliver between 25 and 30 W of power per kilogram of body mass, as we saw in Fig. 2.10, so a 200-lb. hockey player might produce 2,500 W for short periods of time. In baseball, pitchers produce a similar amount—about 3 hp—when throwing a fastball.[6] Because it takes 20 lbs. of muscle to produce 1 hp, the large muscles in the legs and thorax contribute greatly to the power an athlete can generate.

Just like in the slap shot, in a wrist shot the stick often bends as the shooter pushes down against the ice as well as on the puck. This accumulates additional potential energy that is released at the end of the sweep. Because part of the power generated comes from the rebound of the shaft, the player can push the puck at a greater speed

6. R. K. Adair, *The Physics of Baseball* (New York: HarperCollins, 1995), is interesting for its treatment of the mechanics of swinging bats and the aerodynamics of a baseball.

using less power. The ideal amount of flexure the stick allows depends on the shooter's style and strength.

Another popular style of wrist shooting is the slap-wrist shot, a cross between the slap and the wrist shot. It is often used by players who want speed and accuracy as well as quick release. As a combination shot, it blends a short wind-up motion with good puck control made possible by the lower impact velocity. It is usually used at short range, especially when a player is closing in on a goaltender.

Made famous by Montreal forward Maurice Richard, the backhand shot is an effective weapon when used properly. Because the player doesn't face the net, a backhander can take a goalie totally by surprise even though the puck doesn't come at the net as fast as a slap shot or wrist shot. Like a forward shot, the backhander can be swept, hit, or lifted high. Being able to raise the puck with a backhander is handy on a breakaway, as it gives one the option of going around the goalie on either side. Otherwise, it would be easy to predict which way the player will shoot. Despite its usefulness, experts say that quite a few professional players don't master this type of shooting.

The Stick

Considering the force of impact on a puck, it's amazing that hockey sticks don't routinely shatter during a hard slap shot. They are designed to withstand over a hundred pounds of bending, so good hockey sticks are flexible and strong. Yet a few still break during every game, so NHL players keep a couple of spares on hand to be safe. Breakage is usually due to wear and tear taking its toll on a stick, not a single blow. Most sticks break after several hard shots, but the amount of abuse a stick can take depends on its quality (and, therefore, its price).

When scientists and engineers study the elasticity and flexibility of materials, they refer to a material's so-called *Young's modulus,* named after Thomas Young, an English clergyman and physicist who, outside his sermons and experiments, also had an interest in Egyptian hieroglyphs. One standard technique for measuring elasticity is to take a uniform rectangular bar of length l, supported at both ends, and apply a force F in the middle. The flexure s, or the distance over which the bar sags at the middle, is related to Young's modulus E via the formula $s = Fl^3/4Ea^3b$, where a and b are the thickness and width of the bar. With this formula we can estimate the flexure of a hockey stick during a shot.

Young's modulus of oak, the kind of hard wood used to make hockey sticks, is $E = 1.3 \times 10^{10}$Pa.[7] Fiberglass sticks should have a similar modulus, since they have more or less the same flexibility as wood sticks. To see if our earlier estimates of the force of impact during a slap shot are correct, we will now use the flexion of the stick as an indicator for the force on the blade. In Fig. 3.6, Michael Peca's stick has a flexure of about 15 cm. Taking $l = 1.5$ m, $a = 2$ cm, and $b = 3$ cm, we can calculate that the force applied by his arm at midshaft is 600 N. We should keep in mind that the end of the stick experiences a force half that at the middle (a property of the torque), so, in reality, we have an impact force of about 300 N on the ice and the puck. This number agrees well with our earlier estimate for a slap shot in the 70–90 mph range.

Hockey sticks have traditionally been made of solid wood, but not just any kind of wood. Because hardness, resistance, and flexibility are crucial properties, softwoods are ruled out. The best kind of wood is elm, especially Rock elm, a variety of hard, heavy, and resistant tree found in southern Ontario and the eastern United States. Rock elm doesn't split easily but bends very well. Sadly, the recent outbreak of Dutch elm disease, caused by a fungus, has devastated the elm population to a point where the species is no longer a commercially viable source of wood. Millions of trees have been lost because of the microscopic parasite, which blocks the vascular conducts of the trunk and deprives the leaves of water and nutrients. On a brighter side, Dr. Martin Hubbes, a researcher at the University of Toronto, recently developed a treatment to help increase trees' natural resistance to the disease, but at this point we can only hope that research in this area will save the elm population.

In the meantime, the disappearance of elm trees sent the hockey industry searching for alternative materials. Hence the arrival of aluminum and other high-tech materials such as fiberglass, graphite, and even Kevlar, the material used to make bulletproof vests. Although

---

7. *Physics Vade Mecum* (New York: AIP Press, 1981).

solid wood sticks are no longer the norm, reinforced wood with laminated fiberglass or graphite on the exterior is still a very popular combination. Sticks made of nonreinforced wood flex more than other sticks and are too easy to break; lamination makes the wood more rigid. The thickness of the laminate and the type of resin applied on both sides of the shaft determine the overall stiffness. There's a crude rating system in the industry referring to how many pounds of weight are required to bend the shaft. Thus, it takes 100 pounds to appreciably bend Easton's new Synergy 100 (a model I'll come back to later).

Reinforced wood sticks tend to be among the heaviest and usually range from the upper 300- to lower 400-gram area. Their flexibility also changes over time as the wood weakens—a process called fatigue. Hollow metal shafts are a viable solution to the problem of consistency.

The vast majority of metal sticks are aluminum-based. Aluminum has the advantage of being both resistant and one of the lightest metals around. Some sticks are made with the same variety of high-resistance aluminum used to build aircraft. Metal sticks come in two sections: a removable wooden blade and a metal shaft. To keep the stick light, the shaft is not made of bulk metal but rather fashioned like a hollow tube around a light wooden core. They are lighter than wood sticks and are typically in the mid- to upper 300-gram range. Aluminum sticks don't break easily and seem to keep their stiffness, but they will gradually bend with time. Consistency is what makes aluminum sticks so popular. They do have a drawback however: some players don't like them because they can't "feel" the puck as well as they can with wooden sticks.

The newest kids on the block are a whole slew of composite sticks made with materials such as graphite, fiberglass, Kevlar, and resin. They are among the lightest sticks around and usually weigh between 250 and 350 grams. Composite sticks offer a good combination of resistance, flexibility, and durability—qualities reflected in their heftier price. Take for instance the much-hyped Synergy hockey stick by Easton. At $150, the graphite and carbon composite stick promises slap shots that are more accurate and up to 10 percent faster. Many NHL players, including Mats Sundin, have adopted the stick with enthusiasm and claim it has had a positive impact on their shooting. The key is that the well-balanced, one-piece Synergy stick tends to flex more toward the bottom of the shaft, closer to the blade, so that more of the bending energy is released back to the puck instead of being wasted elsewhere.

Companies now produce so many different hockey sticks that there seem to be as many models and styles as there are professional players. Finding the stick properties that are optimal for one's style may be a difficult task, but it is mainly a matter of personal taste. There are no scientific formulas that describe the ultimate "best stick." Some hockey buffs say that a stiffer stick—one with a higher Young's modulus—is better for a slap shot because it delivers a more powerful blow. Although stiffness influences the duration of contact with the puck—just like a strong spring releases a mass quicker than a weak one—the energy comes from the player and not the stick itself. In theory, the same amount of energy can be transferred with sticks of different stiffness, as long as the puck remains in contact with the blade during the entire time the stick is bent. However, excessive flexure can also hurt the accuracy of a shot.

How about the curvature of the blade? Is that important? Many hockey fans have wondered which is better, a straight or a curved blade. Offhand, most of us might say the latter. After all, every NHL player, including goalies, now uses a curved blade. It turns out there are some good physics-based arguments for this choice. During the impact of a slap shot, a straight blade will bend forward slightly because of its elasticity. The amount the blade bends will depend on where the puck makes contact with it. The degree of bending will then influence the angle at which the puck leaves the stick, which impedes accuracy. If the blade is already curved, the bending caused by the blow itself is less important. Wrist shots also benefit from a curved blade, because during the sweep the puck will naturally roll toward the bottom of the curve and leave from the same point every time, thereby ensuring a more consistent shot. On the other hand, remember that the opposite is true for a backhand shot or pass, which will suffer if the blade is excessively curved. This is precisely why professional players shunned the curved blade in the beginning, pointing out that it would make reverse shooting less accurate. The curved stick gained acceptance with the popularization of the slap shot, after stars like Bobby Hull adopted it. He proved that someone using a curved blade could shoot either way just as well with a little practice.

Shooting Accuracy

A shot on the goal, as the saying goes, is never a bad play in hockey. Even a bad shot gives the team a chance to score. Skating around trying to plan the perfect play, meanwhile, puts you in danger of losing control of the puck. Hockey is very much about guessing and strategy, about evaluating one's chances and making the right decisions, about answering questions like: "Should I take a shot through the defender's legs or should I go around?," "Is it better to take a shot from here or should I risk a pass to my teammate in front of the net, who might be in a better position for scoring?," or "If I shoot through the goalie's pads, do I have a better chance than going for a top corner?" A player on a breakaway is always facing such dilemmas, which tend to end in total elation or frustrated regret. Making the correct decision in a split second is what stars like Wayne Gretzky and Mario Lemieux could do so much better than the average player. Yet, because the game involves 12 players, each with their own unpredictability, even the greatest players sometimes make the wrong decisions. We've all heard dejected commentators say things like "He made one pass too many!"

An absolute necessity to good shooting, accuracy is one of the most important elements in hockey. Sure, the puck has to go fast, but speed doesn't help if it goes in the wrong direction. A good part of training is geared toward developing and maintaining good shooting skills. But like throwing darts at a board, even the simple act of shooting a puck at the net has an element of chance in it. The annual shooting accuracy competition, held during the All Star game weekend, seems like a simple event: a player passes a puck to a shooter in front of the net who then tries to hit one of four circular targets, each one foot in diameter, placed at the four corners of the net. He has to do this in as few shots as possible from a distance of about 20 feet, and there is no player or goaltender to block him. Typically, in such a low-pressure situation, the winner takes the honors after firing only five or six pucks. Yet it's interesting to see how many skilled hockey stars will fire two, three, or four pucks in a row without hitting a single target. The competition is a telling example of how much luck and how much skill is involved in shooting. So how is it that some of those not-so-accurate star players still manage to top the point-scoring list at the end of a season? Part of the reason is that they often take shots near the net rather than shoot at small openings. Frankly, they often rely on chance, hoping that the puck will find its way to the net.

Shooting accuracy is easily dealt with using physics. It is statistical in nature and can be described by the theory of probability, a branch of mathematics over a century old. Statistical physics theories have been widely used to describe phenomena in thermodynamics and nuclear physics. Even though each atom cannot be tracked individually, the behavior of the system as a whole can be described very well with probability calculus. Likewise, even though it's impossible to know with certainty if one of Brett Hull's shots will score, predictions can be made about what fraction of his shots will be successful based on his shooting skills.

To see what real-life shooting is all about, let's take former Colorado defenseman Ray Bourque as an example. Suppose he shoots at the net from the side of the blue line, some 70 feet away. The rushing puck finds its way through several sticks and legs before squeezing between the goalie's pads, which are a mere five inches apart. The crowd roars with delight (only if it's a home game, of course) and the sportscasters tell the TV audience about how accurate and powerful the shot was. But what was Bourque's real margin of error? By *margin of error*, I mean the horizontal angle we'll call $\Delta\theta_x$ within which the puck will find the net. We can find it using simple trigonometry: the relationship is $\tan(\Delta\theta_x/2) = \Delta x/2d$, where Δx is the width of the target and d is the shooting distance. Rearranging the equation gives the following horizontal margin of error:

$$\Delta\theta_x = 2\arctan\frac{\Delta x}{2d}. \qquad (3.12)$$

In Bourque's play, the window of opportunity was a tiny 0.3°! This is much smaller than the 6° aperture available when shooting into an empty net from the blue line. Evidently, there was an element of luck in this hypothetical shot, because although Bourque is a highly accurate shooter—he has won eight shooting competitions in the past—no hockey player, no matter how skilled, is able to routinely hit a fast puck with a 0.3° margin of error. That doesn't mean accuracy is not important, otherwise anyone who could skate might join the

NHL. As with any sport, precision is a major factor in the long run, influencing the average over many shots. The more accurate the shooter, the more goals he will score in the end.

When goaltenders Ron Hextall and Martin Brodeur scored overtime empty-netters (shots when the goalie is pulled and replaced by a sixth attacker) in 1989 and 2000, respectively, the margin of error was also very small: only 2°. This is why it took several attempts before they could light up the red bulb. Understanding his luck, when Hextall was interviewed after the game on why the puck skimmed the inner side of the post before entering the net, he jokingly said that was exactly where he'd wanted it to go.

The next step for us is to translate an angular margin of error for a particular situation into a probability of success given a shooter's ability. We now need to know how consistent (accurate) the shooter is. For instance, we wouldn't expect a B-player to have as much of a chance as Bourque on his imaginary goal. The more consistently a player is able to hit a target at a certain distance, such as is measured during a skill competition, the more accurate he or she is. We can't judge a player on a single shot—we need to look at several attempts. A typical shooter at the skill competition will miss the target once every three shots (an estimate, of course, but it's in the right ballpark). This equates to a 70 percent success rate with an angular margin of error of about 3°. On his shot from the blue line, the chance for Bourque of scoring was then

$$\text{scoring chance} = .70 \times \frac{0.3°}{3°} = .07.$$

I should point out that the opening width Δx we use in the previous formula is really the width of the target as it *appears* from the puck's point of view. When a target of width w is seen at an angle β, then the apparent width is then $\Delta x = w \cos \beta$. In other words, there's no window of opportunity for a target rotated by 90°—you can't score if you're shooting from along the goal line. Hence when a player shoots from the far side of the net, the chances of scoring are greatly limited.

So far we have dealt with the simplest scenario, namely shots taken along the ice where there's only one degree of freedom, the horizontal shooting direction. In a three-dimensional world there are

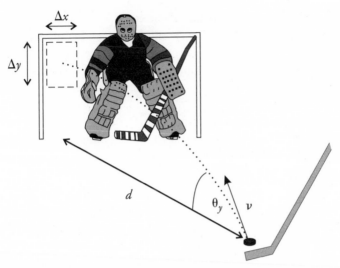

Figure 3.8. When taking a shot at an opening in the net, a shooter has a small margin of error, determined by the size of the opening and the distance of the target. In general, there are three variables involved: the shooting velocity v, the shooting angle θ_y and the direction θ_x. Δx and Δy are the width and height of the rectangular opening.

three parameters to take into account. As Fig. 3.8 illustrates, these are the shooting velocity v and the horizontal and vertical shooting angles (called θ_x and θ_y, respectively). Each one has its own margin of error, related to the size and distance of the target, and we can work them out by going back to Equations 3.1 and 3.2. Given a target of height Δy and width Δx (the target being a net opening), the problem is to find the windows of v, θ_x, and θ_y within which scoring is possible. Equation 3.12 is still valid for θ_x because the horizontal path of the puck is the same whether or not the puck is lifted in the air, so we will leave it at that. Using a few calculus tricks (the calculations are given in Appendix 5), we obtain the following:

$$\Delta v = \frac{\Delta y v^3}{g d^2} \tag{3.13}$$

$$\Delta \theta_y = \frac{\Delta y}{d}, \tag{3.14}$$

with $\Delta\theta_y$ expressed in radians. These equations illustrate a few important points. The larger the shooting velocity, the more room for maneuvering we have on it. This makes sense: the less time the puck takes to reach the net, the less influence gravity will have on its vertical position. Distance, on the other hand, shrinks the margin of error for both v and θ_y.

Remember that for a successful shot, v and θ_y are not independent and cannot take any values we wish. The combination of shooting velocity and aiming angle must be such that the puck has a chance to reach the target without bouncing off the ice first.

An example will give us a feel for how small such margins of error are. A player at the blue line sees an opening of one foot by one foot at the top corner of the net and aims an 80 mph slap shot at it. The shot angle is around 5°, the correct angle needed to hit the target dead center. The player will score within a 10 mph window of velocities and a 1° margin for θ_y and θ_x. If the shot is fired in the correct direction, any speed from 75 mph to 85 mph and any shooting angle between 4.5° and 5.5° will get the puck into the net. In other words, the puck velocity has to be accurate to within about ± 6 percent and the angle to within ± 10 percent.

On a final note, although air friction affects the puck trajectory, it is not expected to significantly change the margins of error, which are approximations in the first place. The conclusions we've drawn are therefore valid with or without air friction.

Chapter 4

COLLISIONS AND PROTECTIVE GEAR

"Let's get ready to rumble!!!" yells the host at the beginning of a World Wrestling Federation show. Considering the reputation of many hockey players, the NHL could use the same opening line instead of playing the national anthems. But unlike wrestlers, hockey players are not trained actors: the hits are real and meant to hurt. If you have seen the *Rock'em Sock'em* video series hosted by the flamboyant hockey commentator Don Cherry, you know that hockey is more like a collision derby than a contact sport. Checking opponents is just as much a part of the game as passing pucks and scoring goals. A good body check is an effective way to gain control of the game by making the opponent cough up the puck. What's more, with an adversary out of play and recovering from a blow, the overall ability of the other team is affected.

Hockey is by no means the only violent team sport. Football and rugby—just to name two—are just as rough. But there is a distinctive spirit that makes hockey stand out from the rest. You would be hard-pressed to find another sport in which the punishment for engaging in a fistfight is a mere ten minutes in the penalty box. In comparison to the way other sports deal with belligerence, this is only a slap on the wrist. This culture of toughness in hockey has produced a class of players called "enforcers," "goons," or, in Quebec, *les hommes forts,* who are nothing short of NHL-sponsored boxers on skates. Long before basebrawls (as commentators call fights during baseball games) became a common occurrence in the pro leagues, hockey had already

Figure 4.1. A good clean body check can stop a dangerous opponent in his tracks. Here Pittsburgh Penguin Hans Jonsson shows Ottawa Senator Marian Hossa the fine art of gaining control of a puck. CP Picture Archive (Fred Chartrand).

established a tradition of fighting. During the playoff series between the Quebec Nordiques and the Montreal Canadiens (the hockey equivalent of baseball's "subway series" in New York) in the 1980s, fans accepted bench-clearing brawls as an integral part of the show. A decade earlier, the Philadelphia Flyers had helped establish a legacy of roughness in the NHL. Nicknamed the "Broad Street Bullies" (a title inspired by the location of their arena), the team played a style of hockey that didn't quite fit with their hometown motto, the City of Brotherly Love. Nonetheless, their resilience earned them back-to-back Stanley Cups in 1973 and 1974.

To promote sportsmanlike behavior the NHL introduced an award, the Lady Bing Trophy, to be given each year to the player who combined the best gentlemanly conduct with a high standard of play. Although the trophy has been awarded to superstars like Wayne Gretzky, it has done little to change the mentality of the game. Like it or not, the Lady Bing is not the prize most hockey players dream of winning. The fact that two-time Lady Bing winner Paul Kariya was recently suspended for slashing an opponent indicates how rough spots during the game can infuriate even its most gentlemanly players.

Fortunately, most of the usual roughness in hockey is without grave consequences. When tempers flare and fights break out, the participants seldom get really hurt. Debilitating checks, collisions, and slashes do not happen often, thanks to well-designed protective equipment. Though many critics (some of whom have never played the game) would have us believe otherwise, the seemingly ruthless violence in hockey is controlled in some way. Players don't check indiscriminately. In addition to the possibility of getting injured, there is the danger of putting oneself out of position or missing a scoring opportunity. There are also rules—some of them unwritten—by which all players must abide if they don't want to end up in the penalty box or hear the disapproval of every hockey fan in the stadium. Hits from behind are not tolerated, players must keep their stick and elbows down during a check, and referees will intervene in a fight as soon as one player has fallen. To discourage fighting, the NHL rulebook states that the instigator—the one who drops his gloves and throws the first punch—gets an extra minor penalty, meaning his team will play short a man for two minutes. This code of conduct plays an important role in hockey; from a Darwinian point of view, the sport would never have survived and evolved into what it is today if too many injuries had been allowed to happen.

This chapter explores the physics of hits and collisions. Understanding what goes on during a heavy blow gives us insights on how to build proper equipment and prevent injuries. Our goal here is twofold: to study collisions—their impact and how they cause injuries—and to look at the role of equipment designed to prevent such injuries.

Let's Dissipate Energy!

The concept of energy is an all-important one in physics. It tells us how much work has been or can be done, regardless of the intricate details of the process. In a way, the concept of energy simplifies life greatly. When the electricity bill states we have consumed 1,000 kWh of energy over a month, it is telling us how much work the electrical appliances have done. Just like money is valuable because of its purchasing power, energy commands a price because of its work-producing power.

In previous chapters we dealt with the idea of kinetic energy, the energy associated with a moving mass. When a force is applied to move a body, it accomplishes work. As the body accelerates (or decelerates), the work is transformed into kinetic energy, given by $K = 1/2mv^2$. This principle is called the "kinetic energy theorem" and is a useful tool in dealing with many problems in physics. Conversely, the kinetic energy of a moving mass can be harnessed to produce work, which does not necessarily mean beneficial work—take the damage resulting from a car crash, for instance. Similarly, an athlete moving on ice carries with him enough energy to inflict injury both on himself and on somebody else. According to the definition of kinetic energy, the heavier and faster the player is, the larger his kinetic energy will be.

Kinetic energy can be transformed (or dissipated) into other forms of energy, such as heat, or be used to permanently deform and break things. A smooth way for a hockey player to dissipate his kinetic energy is to slow down by braking. In the process, the friction force of the ice converts the player's kinetic energy into heat and the force used to break off ice chips. A more dramatic way to waste energy is to collide with someone else. When two players traveling in opposite directions smash at mid-ice (or in the "death zone," as it is sometimes called), the fact that they suddenly stop means their kinetic energy has gone somewhere.

Varying as the velocity squared, kinetic energy is more affected by a player's velocity than his or her mass. Thus, the energy of a hockey player going at breakneck speed is huge. Take for instance an NHL athlete of average weight skating at an average speed, say three-quarters of his top speed. The typical guy on the 2000–01 New Jersey Devils roster (not including goaltenders) weighed 205 lbs., or 93 kg. Though this seems big, the Devils are pretty much in line with the rest of NHL teams. The highest skating speeds recorded in hockey are of the order of 30 mph, so it's safe to say that a player could commonly reach 75 percent of that, or 23 mph (10 m/s). (In fact, a fast skater will complete a lap around the rink in 13.8 to 14.5 s when starting from rest. That's an *average* speed of 25 mph.) At 23 mph, our 205-lb. Devil carries 4,700 J of kinetic energy. When two such players collide and come to a stop abruptly, 9,400 J of energy will be dissipated, which is enough to power a 100 W light bulb for a minute and a half! We now understand why participants might "see stars" after such a collision.

This sounds like an awful lot of energy, but how does it compare with that of other contact sports? Football players are significantly bigger, but they don't move as fast. The fastest runners on the field— the running backs and wide receivers—can cover a 40-yard stretch in about 4.5 seconds when fully equipped, or in less time if they aren't wearing gear. This corresponds to a respectable average speed of 8.2 m/s, or 18 mph. The average player on the 2000–01 San Francisco Forty-niners tips the scale at 242 lbs., or 110 kg (not accounting for the quarterbacks and kickers). Therefore, a running back going as fast as he could would carry a mere 3,700 joules of kinetic energy. Bigger football players (those linebackers who weigh up to 350 lbs.) run much slower and have about the same or less kinetic energy than running backs. So a lightweight hockey player like Paul Kariya, at 180 lbs. but skating at full speed, has more kinetic energy than an offensive lineman! Of course, we are neglecting the weight of his equipment here, but that is typically only a small fraction of the hockey player's overall weight—about 15 percent.

There are sports where even more kinetic energy is at play. When running at full throttle—near 12 m/s—in a 100-meter race, the 79-kg Olympic gold-medalist Maurice Greene carries 5,700 joules of energy. The fastest sport on Earth (apart from skydiving) is downhill skiing, where speeds in excess of 230 km/h (or 65 m/s) have been reached according to the *Guinness Book of World Records.* An 80-kg skier going that fast has 170,000 Joules of energy, enough to cause a huge amount of damage in a fall (or power a light bulb for an entire half-hour, not just a few minutes). This kind of kinetic energy dissipation is along the lines of that found in a car crash. But even then, car speeds are typically only half that of downhill skiing. Fortunately, neither sprinting nor downhill skiing are contact sports (not usually, at least).

In a collision or a fall, kinetic energy is dissipated through different channels. Heat, noise, and other vibrations are always produced. Sometimes permanent deformations occur: broken bones, sprained joints, or torn ligaments can end careers. The amount of damage incurred from a body check depends on a number of factors, one of them being the weight of the players involved.

Big versus Small—Who Wins?

The player that gives a body check does not always come out the winner. Answering the question of who may benefit the least in a heavy collision is easy for anyone who has played full-contact hockey. Most players with common sense would rather clash with a featherweight than with a colossus on skates. To understand to what extent body weight plays a role in checking, we need to look at the physics of collisions in general.

Studying collisions is greatly simplified if we introduce the notion of momentum, defined as

$$p = mv. \tag{4.1}$$

Note that p is the momentum due to linear motion, as opposed to the circular momentum we examined in the previous chapter. Momentum and kinetic energy are somewhat related, though not equivalent. Whereas energy might not be conserved during a collision, momentum always is. This is a direct consequence of Newton's laws of mechanics.

In physics, a collision is defined as a strong interaction between two bodies that happens within a very short period of time. The force of impact is much greater than all the other forces applied on the colliding bodies (friction, gravity, etc.), so they can be neglected. This is a valid assumption even for hockey players, as we will see later. One interesting property of collisions is a direct consequence of Newton's third law. Because the force on each player is equal and opposite in direction, they will accelerate in opposite directions until they are no longer in contact or until they are moving together with the same velocity. (I should emphasize that they *accelerate*—or decelerate—but don't necessarily *move* in opposite directions.) As a consequence of that, each hockey player has the same change in momentum, but with opposite signs, so that the total momentum of the two players together remains constant throughout the collision. Mathematically it looks like:

$$m_1 v_{1i} + m_2 v_{2i} = m_1 v_{1f} + m_2 v_{2f}, \qquad (4.2)$$

with subscripted 1 and 2 referring to each player's mass and i and f referring to the initial and final velocities.

I should point out that we are dealing with collisions in one dimension only. Perhaps a more realistic scenario for hockey would be to consider the two-dimensional case, since players move on a surface, not a fixed rail. However, the equations for momentum conservation in two dimensions is exactly the same: we simply write one equation for the x components of the velocities and one for y. Take Fig. 4.2a as an example. We choose x to be the line of motion of player 2, in which case the relevant velocities are $v_{1i} = v_1 \cos\theta$ for player 1 and $v_{2i} = v_2$ for player 2. Along y, they are $v_{1i} = v_1 \sin\theta$ and $v_{2i} = 0$. The force of impact will be distributed along both the x and y axis. Against the board (Fig. 4.2b), the relevant velocity components are

(a)

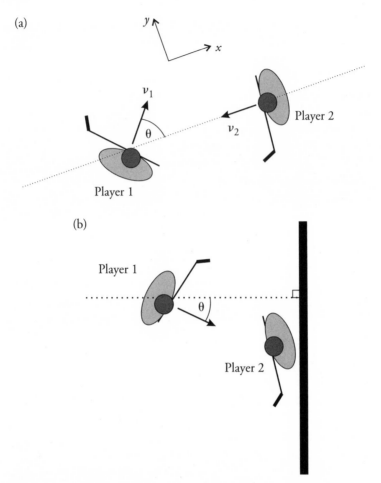

Figure 4.2. (a) When players collide at mid-ice, the relevant velocity components are along the line of motion of either player, in this case player 2. (b) Against the board, the velocity components along the board do not matter as much as the perpendicular ones.

perpendicular to the board: they are $v_{1i} = v_1 \cos\theta$ and $v_{2i} = 0$. In both situations, if the angle θ is small, most of the impact force will be along the dotted line, and Equation 4.2 alone describes the collision very well.

To go any further, we need more than just Equation 4.2, otherwise, even when we know the mass and the initial velocity of each body, we are left with two unresolved final velocities. In some collisions, like that of billiard balls on a pool table, there is no loss of kinetic energy. When they impact, they compress slightly (though not enough to be visible to the naked eye) and spring back to completely restore their kinetic energy. The same thing happens when a good rubber ball bounces off the wall. For such collisions, we can write an additional equation for the conservation of kinetic energy and find the final velocity of both colliding objects. Alas, the same cannot be said of hockey players. When they smash into one another, they produce inelastic collisions in which energy is absorbed. Another example of inelastic collision is when one car rear-ends another and the two get stuck together. Perfectly inelastic collisions—when two objects become entangled and move together after they collide—are the simplest to deal with because one final velocity variable is eliminated. So, even though kinetic energy is not conserved, if the collision is perfectly inelastic we can still deal with the problem using Equation 4.2 alone.

In hockey, there is little bouncing after checking on the board or a mid-ice collision. We don't see hockey players rebounding off one another like billiard balls, although this is quite interesting to imagine. Though flexible, the human body is not particularly elastic, and much of the impact energy is absorbed by the body tissues. To my knowledge, no one has measured the "bounciness" of the human body, but, looking at hockey players colliding, it is clear that their final velocities are much smaller than their initial ones. And, of course, athletes are not punctual masses but three-dimensional bodies that flex, stretch, and rotate. After colliding, they may push each other away and fall backward, or, if one player sees the attacker coming and lies low, the other one might roll over or spin. All sorts of complex movements are possible.

Because each situation is different, the analysis that follows is an approximation. We will assume the final velocity of both athletes is the same—in other words, it's a perfectly inelastic collision. This allows us to use $v_{1f} = v_{2f} = v_f$ and reduce Equation 4.2 into

$$v_f = \frac{m_1 v_{1i} + m_2 v_{2i}}{m_1 + m_2}. \tag{4.3}$$

Figure 4.3. Mid-ice collisions, like this one between Vancouver's Pavel Bure and Detroit's Kris Draper, tend to be more severe than board checks because more kinetic energy is involved as both players move in opposite directions. After colliding, each player's final velocity—though not null—is typically much smaller than his initial one. CP Picture Archive (Chuck Stoody).

Now that we know v_f, we can estimate the total loss of kinetic energy ΔK sustained by both players:

$$\Delta K = \frac{1}{2}\, m_1 v_{1i}^2 + \frac{1}{2} m_2 v_{2i}^2 - \frac{1}{2}\, (m_1 + m_2) v_f^2. \qquad (4.4)$$

To get an appreciation for how much energy gets dissipated, we'll use a realistic scenario. Eric Lindros (236 lbs.) is skating at a modest speed of 4 m/s and crashes into a stationary Ed Belfour (182 lbs.) standing a bit too far from his net. (Actually, this action would be a punishable offence called "charging the netkeeper" according to the NHL rulebook, but we'll use it anyway as an illustration.) From Equations 4.3 and 4.4, we find the final velocity of both players is 2.3 m/s and the energy loss is 360 J, which is enough to power our 60W light bulb for 6 seconds. This energy is dissipated through both players, and, even though the force of impact on each is the same, the consequences are very different. The lighter player will suffer a greater acceleration. Indeed, using Newton's law $F = ma$ (or $a = F/m$), we conclude that during the impact $a_1/a_2 = m_2/m_1$. This means Belfour will accelerate at a rate 30 percent faster than Lindros as a result of the crash.

Impact force and acceleration are two important elements when it comes to injuries. The greater the force of impact, the larger the pressure on the shoulders and the rib cage. The faster the acceleration, the greater the impact internal organs (including the brain) will suffer and the more shaken-up the player will be. Therefore, a lighter player is more likely to come out the loser after a heavy check, just as common sense would have us believe. This is a growing problem for the NHL today because players are now significantly heavier than before. As a result, the disparity between smaller and larger players is widening with time, which increases the likelihood of severe injuries. Twenty years ago, athletes weighting more than 220 pounds were few; today, each team has a couple of heavyweights above that barrier, and some of them are even good scorers.

Fig. 4.4 shows the change in weight and height of the average NHL player over the last three decades. If this trend remains the same, extrapolation on the data shows that the average NHL player would stand 1.90 m and weigh 98 kg (6'3" and 216 lbs) by 2025.

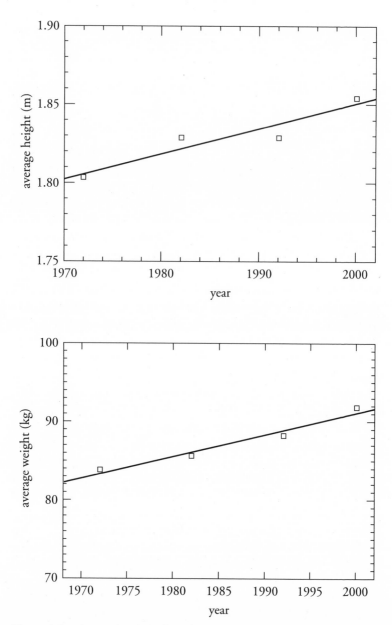

Figure 4.4. Average height and weight of NHL players over the last three decades. The data points were recently released on a hockey newscast on the TV station TQS in Montreal.

On top of being heavier and taller, modern hockey players are better trained, faster on their skates, and, by the same token, more dangerous than ever. Even though good things come in small packages, as the saying goes, some lighter players who have enough talent to become superstars have a harder time thriving in the big league. It became evident that even the Great One, Wayne Gretzky, suffered from the extra strain of playing against younger, bigger guys later in his career. This comes as no surprise, considering that NHL players had gained on average 15 lbs. over the span of his career. Another example is Saku Koivu, the promising but small Montreal Canadiens captain, who has missed over a hundred games in the last four seasons because of injuries. Players like Theo Fleury, at 5'6" and 180 lbs., are the exception rather than the norm nowadays. Because of the increasing player size in the NHL, some junior hockey players are just too small to be drafted in the league, and consequently some of the best talent is wasted.

The Force of Impact

The amount of energy dissipated during a body check is only an indication of the potential damage that can be done. The *force* of impact is what matters most. After all, our bodies don't measure kinetic energy, but we sure can feel a good knock! So how do we apply physics to estimate the force of impact between two players?

It is important to realize that the force of impact during a hit is not constant but varies greatly within the duration of the collision itself. At the beginning, the force is very small, but, as players smash closer together, it increases dramatically and reaches a maximum. Then it drops eventually and returns to zero as they move apart. The same fluctuation goes on in all types of collisions, whether we are talking about a stick hitting a puck, a club driving a golf ball, or a bat hitting a baseball. Trying to find the force at any given time is a complex problem, so we will be working with the *average* force of impact. In doing so, we are missing bits of information, in particular the maximum force—that at the peak of the collision. We just need to keep in mind that this maximum is quite a bit larger than the average.

To find the average force of impact F in a collision we need two parameters: the energy dissipated and the total length of deformation d of the objects during the impact. In a way, d is the equivalent of "crumple zones" in cars—areas designed to absorb the shock by collapsing. From the theorem of kinetic energy, we conclude that the work done by the force is Fd and is equal to the total loss of kinetic energy, so that $\Delta K = Fd$, which we rearrange into:

$$F = \frac{\Delta K}{d}. \tag{4.5}$$

It is clear that rigidity will play an important role here, because the stiffer the colliding bodies are, the smaller the deformation and the larger F will be. Don Cherry, the hockey commentator for the CBC in Toronto, who has absolutely no background in physics, had it right when he said: "The reason so many injuries happen in hockey today is because guys are more muscular and stiff. They don't give!" The primary role of the protective layers of equipment and the natural "body cushion" of players (the skin, fat, and muscle tissues) is to help reduce the shock by increasing a player's deformation thickness d. We can easily imagine that receiving a body check while wearing a thick foam padding wouldn't feel the same as colliding with no protection at all. For the same physical reasons, it is easier to land on a trampoline than on a hard concrete floor. The trampoline sinks by a fair amount, whereas on concrete, our legs are the only objects that move to absorb the shock.

We have already estimated the energy dissipated by colliding hockey players. Now the problem is finding their deformation length, or "crumple zone." This is rather difficult because we are dealing with a complicated structure that has many moving parts. In the case of simple objects like billiard balls, it's easy to find d based on the stiffness of the bulk material from which they are made. Obviously, for a hockey player, d is never the same from one check to another. There are many layers involved in the compression: outer padding, fat and muscle tissues, and bone structures. Though rigid, parts of the skeleton (especially the rib cage) may indeed provide some compression. Other bones, like the collarbone, which protects the thorax from excessive lateral stress, are less yielding. They are often the first ones to break upon a violent sideways impact. (Collarbone injuries are especially prevalent among younger hockey players whose bodies are not fully developed.)

Deformation also varies to some extent with the speed of impact. The more violently players collide, the more their bodies will compress. Another factor is the way the body check is aimed and whether either player is standing upright. At this point we could venture into making elaborate "body stiffness" models to account for deformation as a function of speed, weight, and other parameters, but the exercise would be futile. For the purpose of this discussion, an educated guess on d will suffice.

We know that players don't flatten by one meter when they check one another, but their bodies probably deform by more than a few centimeters. Considering that the typical width at the shoulder of an adult male is on the order of 50 cm, an NHL player and his equipment will probably compress by as much as 10 cm in a heavy collision. Using $d = 20$ cm for the total deformation length of both players and putting it in Equation 4.5, we can obtain an estimate of the average force of impact in our example involving Lindros and Belfour: 1,800 N, which is equivalent to 400 lbs. of force! That's quite a blow, but don't forget that Lindros is a big guy and the collision force only lasts for a split second.

Ready for Boarding!

Mid-ice collisions may be the easiest to physically analyze, but they are not the most frequent. In fact, they are often accidental and tend to cause more injuries. (I remember seeing a Quebec Nordiques player break both his legs this way.) Instead, the great majority of collisions occur along the board, where players use body checks to gain control of the puck. Fig. 4.5 illustrates the tactic convincingly. Sometimes the only way to get past a defender is to go through the "wringer"—that is, squeeze between the opponent and the board and hope to come out on the other side clean and dry!

The board is nothing but a wall encircling the ice that meets a number of requirements. NHL arenas must have a board made of wood (painted white) or fiberglass that is at least 40 inches (1.02

Figure 4.5.   Ready for boarding? Ottawa Senator Todd White doesn't think so . . . CP Picture Archive (Tom Hanson).

meters) and no more than 48 inches (1.22 meters) above the ice surface. The suggested height is 42 inches. It must be void of ob-structions that could influence the play (for example, doors need to open away from the ice). A yellow kick plate covers the bottom of the board, and glass windows are placed above it. The glass allows spectators to see the action while also protecting the crowd from flying pucks. Just like car windows, these are designed to shatter and crumble in small pieces, reducing the danger of flying shards of glass whenever a plate breaks upon impact.

As far as body checking against the board goes, the question is, How does the board influence the force of impact? Does it reduce or increase the shock? At first glance we may be inclined to say that it will increase the impact, because the two players are hitting a wall and they both come to a complete stop, unlike in mid-ice collisions. All the kinetic energy vanishes. On the other hand, if the wall flexes, the blow is absorbed in part by the board (the cushioning length increases). The combined effect of board displacement and total dissipation of kinetic energy determines whether the impact will be lessened or not. What is certain is that the stiffer the board, the greater the impact. Coaches and players sometimes complain about boards in particular arenas that are too stiff.

The amount of flexion the board provides varies from one rink to another, and it also depends on how hard you slam into it. The bending is usually not visible on TV, but if you sit close to the board while at a game, you can see it give a couple of centimeters during a healthy check. In our previous example, if Belfour happened to be standing behind his net and was checked against a rigid board by Lindros (again, a violation that would trigger more than just a penalty in real life), the force of impact would be much higher than that of the same collision at mid-ice: 4,300 N, or 960 lbs., instead of 400 lbs. Both players stop against the board, and all kinetic energy is dissipated. But if the board gives an extra 5 cm to increase the total deformation length (of board and bodies) to 25 cm, the impact would be lessened to 770 lbs., a 25 percent reduction I'm sure Belfour would welcome. In this second scenario, the body check against the board is still more grueling than at mid-ice. But, in general, collisions at mid-ice tend to be more violent and dangerous because *both* players are skating—usually in opposite directions.

Injuries and Safety Gear

Hockey could never have evolved into such a fast-paced sport had not it been for properly designed protective gear. The increasing speed of the game is due in part to the development of high-tech equipment that protects athletes from heavy blows. Don't forget that in addition to body checks, hockey players face the threat of flying pucks, nasty slashes, sharp skates, and a hard ice surface. Unless you want to leave the game on a stretcher, you need to wear the proper gear. As many parents will tell you, hockey is perhaps the most expensive sport for kids to play because of all the equipment involved. It's even worse for goaltenders, who need to spend a few thousand dollars on proper gear.

With physics we can understand the twofold protective action of sport equipment. First, it serves to absorb the shock during impact, and, second, it helps redistribute the blow over a large area. The cushioning effect, as discussed earlier, is simply to reduce the force of impact by providing room for extra deformation. But the amount of force alone is only one side of the story. The area over which that force is applied is equally to blame for injuries. In other words, *pressure* is what matters here. We are all familiar with the concept of pressure, which we hear about every day in weather forecasts, but it is applicable to many other situations. A needle, for instance, easily penetrates the skin because the pressure on the tiny surface area of its tip is huge, even when just a small force is applied. When a hockey player suffers a hit, the pressure of the blow is what determines the mechanical stress on a particular point on the body. During a body check, the impact may be distributed over the shoulders, the hip, and the leg, which helps alleviate the overall pressure. But if one player sticks his elbow out, the blow is concentrated over a small area and the chances of injury are increased. Unfortunately, many players learn to use their elbows and pull other dirty tricks as defense tactics.

One role of protective equipment is therefore to reduce the pressure of impact, especially against objects with smaller surface areas—like stick edges and pucks. A slap shot at 80 mph sends a puck flying with 110 J of kinetic energy. Without equipment, being hit anywhere on the body by a puck moving that fast inevitably means injury, as the shock occurs over an area of only a few square inches. Assuming the puck stops after traveling 2 cm into the body tissues, and taking the contact area to be about 2 square inches, we obtain 40 times the atmospheric pressure! Fortunately, a good kneepad will distribute the force over a much larger area (roughly ten times larger) and greatly reduce the damage.

In order to be efficient, protective gear must be designed properly. In a hostile environment like an NHL game—one of 82 games during a full season, not including playoffs—players' gear receives a lot of abuse and must be strong enough to sustain repeated blows. In theory, a simple layer of padding, if thick enough, would protect against all types of dangers. Alas, such all-purpose padding would be so thick the athlete would look more like the Michelin man than a hockey player. Instead, force redistribution is efficiently achieved with a thin but hard plastic shell. Underneath this shell, a thin layer of soft material, such as foam or felt, serves to absorb the energy. This concept of "hard on the outside, soft on the inside" is the basic principle underlying the design for many pieces of equipment, including shoulder pads, elbow pads, gloves, kneepads, and pants. It's also the same idea behind the Kevlar bullet-proof vests worn by police and soldiers.

There are exceptions to the rule, notably goalie pads, which are thick, bulky, and soft throughout. Harder, thinner pads could theoretically do the job, but the puck would rebound off them, possibly back into the path of the forwards, who would then have an easy chance to score. Goalie pads are made of leather and filled with light synthetic fibers that absorb the shock very well and keep the puck from bouncing. Other pieces, such as the chest protector, also act the same way. Of course, wearing such thick padding makes goalies sweat a lot more during a game, which is why a bottle of water is always present on the top of the net.

The wear and tear on hockey equipment is a major issue in the NHL. For safety purposes (and also for comfort), athletes periodically replace items. Many buy new equipment at a rate that would be too costly for amateur hockey players. Former Montreal captain Vincent Damphousse was known to have gone through four pairs of gloves in a month!

Yet sometimes even the best padding is not enough. In 1999, in a game against Chicago, defenseman Al MacInnis fired a slap shot from the blue line, and Jocelyn Thibault stood sharp and ready to stop it. It would have been a great glove-save by the Black Hawks goaltender—if the puck had not perforated his glove, broken his finger, and dribbled into the net. Because the glove is a complex and important component of the goalie's attire, it cannot be too bulky. A good glove offers a fine balance between flexibility and protection. To protect the palm, a thick layer of leather is used at the middle, but in this case it wasn't enough to shield against MacInnis's shot.

Head Injuries

The helmet has allowed mankind to pursue "head-cracking activities," Jerry Seinfeld once joked. Your brain is your most important asset, so no one will disagree that wearing a helmet in hockey is a must. In light of what we know can happen without head protection, it's mind-boggling to think that for many decades NHL hockey players shunned the helmet but wouldn't dare step on the ice without gloves or kneepads. Although the helmet made its appearance on the rink a long time ago, its use didn't spread right away. Goaltenders, as the ones on the receiving end of most high-velocity shots, were the first to embrace the concept of head and face protection. Even so, many goalies thought a face mask would reduce their field of vision and make them look like they were afraid of the puck. It took several unfortunate incidents before someone stood up for the sake of common sense. During a game against the New York Rangers in 1959, Jacques Plante of the Montreal Canadiens was struck in the face by a puck. The impact broke his nose and made a deep cut that required seven stitches. The star goalie, who had won six Stanley Cups during his career, left the game to be patched up and refused to return without a mask. The situation became problematic for coach Toe Blake, because in those days teams did not have spare goalies. After much arguing, Plante was allowed to make history by returning with a face mask, which he never parted with after that day. The crowd gasped in disbelief as he stepped back on the ice. Plante had made the mask based on a modified welder's mask, and he had worn it before, but never in a regular game.

Functional but not very aesthetic, early face masks resemble those seen in horror movies like *Silence of the Lambs* and the many *Friday the Thirteenth* pictures. Nowadays they are much more sophisticated—and expensive—pieces of equipment. High-tech materials like fiberglass and Kevlar provide better resistance. Some masks are molded directly from the goalie's face to produce the most comfortable fit. In pro leagues, masks are usually painted with artwork meant to either embellish or intimidate, like the screaming eagle Ed Belfour's wore during his Chicago days.

Helmets gained acceptance among regular players in the 1960s and '70s, but only on a voluntary basis. In an inevitable move, the NHL mandated headgear in the 1979–80 season for all rookies. Veterans who had joined the league before that year were allowed to continue showing off their hairdo if they signed a legal waiver stating they would not hold the league responsible for head injuries. The era of bareheaded hockey officially ended in 1997 when St. Louis center Craig MacTavish hung up his skates after a 17-year career.

There are many threats to a player's head in a hockey game. Pucks, elbows, and highflying sticks are three of them, but there is also the danger of bumping against other players, crashing into the board, or falling onto the ice. These can cause a type of trauma called a *concussion,* which has made headlines in recent years because of the number of star players who suffer them. Pat Lafontaine was forced to retire early from the New York Rangers, as was Eric Lindros for a while. Mike Modano of the Dallas Stars and Anaheim's Paul Kariya are two oft-cited cases of NHL heroes who fell victim to multiple concussions. Lindros's questionable return during the 2000 playoff series against New Jersey, while he was still recovering from his latest head injury, could have been a career-ending mistake. During a forecheck rush he didn't see New Jersey's Scott Stevens coming and went down hard. This gave him his sixth concussion in two years, at which point doctors said he should retire early—as his younger brother Brett had, a couple of years before—if he didn't want to end up permanently brain-damaged.

In addition to the increased frequency of concussions among professional athletes (especially in football, hockey, and boxing), a series of studies contributed to the rising concern over head injuries. Researchers found that the chance of debilitating injuries increases with the number of blows an athlete suffers. This may seem obvious, but there was another, more important discovery: the effect seems to be not additive but multiplicative. Even if a player feels fine after a few minutes of rest, a second hit to the head is far more likely to have long-lasting or permanent consequences. The brain simply becomes more fragile and susceptible to damage after the first hit. In the old days, coaches and team doctors would send a shaken player back onto the ice as soon as he felt okay, unknowingly exposing him to danger. Today, team doctors make sure the extent of the injury has been properly determined before a player returns to the ice. This makes sense not only on a personal level—no one wants to see a friend get hurt—but also from a business viewpoint. NHL players are, after all, valuable assets in a multimillion-dollar business.

Medically, a concussion is simply a change in mental state following a violent blow to the head. Articles on the subject recently appeared in a special issue of *Scientific American* dedicated to the science and technology of sports.[1] During a concussion, the victim may or may not lose consciousness, but confusion and dizziness always follows for a certain amount of time. Other effects include headaches, disorientation, loss of memory, and blurred vision. In more severe concussions, internal bleeding might occur and blood clots (hematoma) may form around the brain, causing dangerously high levels of pressure.

You don't have to be a star player on the Rangers team to be at risk of getting a concussion. I've had a couple of minor ones myself during my decidedly amateur hockey career. When I was younger, I collided with a player and fell backward onto the ice, hitting my head rather hard. After a few seconds of dizziness, I stood up and reassured my coach that I was fine. I stayed in front of the net and the game went on, but for the remainder of the game I had difficulty recognizing my teammates from my opponents! The color of the uniforms didn't help: I simply couldn't distinguish our maroon jerseys from their green ones. To add to the problem, the motion of the players looked discontinuous, so it was hard to focus on the puck. Luckily, there were no long-lasting consequences (at least I think not), and I had a good excuse for letting in more goals that day. More recently, I collided heavily with a player rushing on a breakaway. There again, my vision was affected for about a half an hour, although to a lesser extent. In both cases, however, I probably should not have stayed on the ice.

---

1. "Building the Elite Athlete," *Scientific American,* special issue (2000): 44.

Physics can help us better understand what goes on when someone suffers a concussion. During a heavy collision, the force of impact causes the head to undergo a large acceleration. Because of inertia, the brain, which weighs 1 to 1.5 kg (brain size does vary a lot from one athlete to another), slams against the skull, resulting in the stretching of neural cells. It's a bit like when the driver in a car accident gets pushed toward the steering wheel after the vehicle has abruptly stopped. The squeezing of the brain causes the concussion. Neurologists tell us that a concussion is not a "bruise on the brain," as some people think, but rather a chemical imbalance that has a cascading effect. Neural tissues are usually physically intact after the blow, but a series of harsh chemical reactions is unleashed, triggered by the simultaneous firing of many neurons upon impact. As with any other physical injury, it will take days, weeks, or even months for the brain to return to a normal state.

The prevalence of head injuries in sports like hockey and football is not blamed solely on the roughness of the activity. Improper headgear is sometimes the culprit. In January 2000, the Canadian Broadcasting Corporation (CBC) ran a documentary showing how helmets—both new and used—failed to meet the accepted standards of protection.[2] In this particular study, none of the five randomly selected hockey helmets (ranging in age from a few years to many years) passed the Canadian Standard Association test, the criterion used to design them in the first place. Although manufacturers guarantee that the helmets will be safe for a period of time ranging from three months to almost three years, amateur hockey players typically wear their helmets far longer than that, not realizing they don't meet the standards anymore.

To understand the difference between a good and a bad helmet, we must look again at the physics of impact. The skull is a rigid bone structure that shields the brain from external blows. But because it doesn't deform easily, it provides very little cushioning. On impact, then, a significant fraction of the energy is absorbed by the soft internal tissues of the brain. At the same colliding velocity, the skull will suffer a greater acceleration than will softer body parts like the shoulder or a hip. Fig 4.6 shows the helmet's function, which is

2. "Hockey Helmet Safety," *Marketplace*, January 11, 2000, Canadian Broadcasting Corporation.

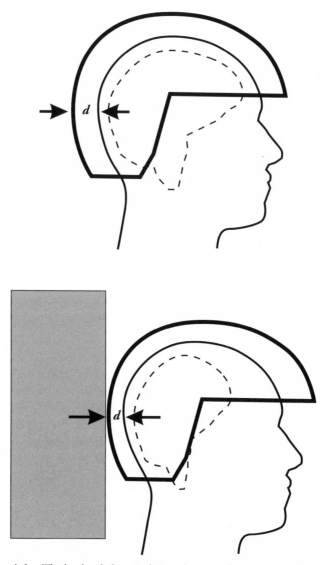

Figure 4.6. The hockey helmet is designed to provide extra internal deformation during a blow to the head. The cushioning distance d reduces the force of impact as well as the acceleration of the skull and the brain, thereby reducing the risks of head injury. Excessive compression of the brain leads to concussion.

to provide adequate deformation room (cushioning) to lessen the acceleration of the head and its contents. In addition to cushioning a blow, the hard plastic shell of the helmet diffuses the pressure over a larger area and reduces the risk of skull fracture.

Based on these facts, the design of a helmet must be carefully thought out in order to be truly effective. Not just any kind of plastic shell with internal foam padding will do. First, the shell must be able to withstand the impact from a free fall of about one meter onto a hard floor without cracking. Older helmets often have weak spots and loose or missing screws, so they are vulnerable to such impact. Second, the inside layer (made of Styrofoam or other soft material) has to have the appropriate stiffness. In the best case, the cushioning layer would be compressed to its maximum only at the peak of the impact, that is, at the moment the head stops moving. If this happens, the deformation is maximized and the head's acceleration has been reduced as much as possible. Of course, no single padding structure provides optimal cushioning for all intensities of impact. The ideal helmet would have a variable-stiffness layer that would harden on heavier impacts. If the material is too hard, it will not compress to its full potential. If the material is too soft, the head will hit the plastic shell and suffer an even greater acceleration. The texture and shape of the cushion is chosen so as to minimize the force of impact for a typical, midrange fall.

Time is an enemy, even to high-quality helmets. The cushion layer becomes worn down over time as it is exposed to a lot of sweat and humidity. The layers shrink and stiffen, and after a time they are unable to provide adequate protection. This holds true for bicycle and motorcycle helmets as well. It is crucial to check or replace safety gear once in a while.

Ironically, although the science of head protection is quite advanced, manufacturers are slow to apply new technologies because of the legal elements involved. They tend to stick with conventional designs because they risk lawsuits if they dare to innovate and stand apart from the crowd. An injured player could successfully sue a company even if his helmet was in fact safer and the new design was supported by research.

Chapter 5

KEEPING THE NET

To anyone who's not a goalie, the task of preventing pucks from entering the net seems pretty simple. That's until they try it for themselves! Just wearing the 50 pounds of bulky equipment and moving around on the ice is awkward enough, let alone blocking an avalanche of pucks while keeping your composure. When I was unable to attend a game, I sometimes let a teammate guard the net. Afterward they told me how much their outlook changed and how weird it felt to stand on the line of fire and try to move quickly while padded like an armored truck.

Another challenge facing goaltenders is the pressure of continually being on the hot spot. Guarding the net at the amateur level might sometimes be just as enjoyable as playing any other position, but this is not the case at the professional level, where the goaltender is as crucial as the pitcher in baseball and the pressure is as fierce. Other hockey players can afford to slack off from time to time without attracting too much attention, but if a goalie loosens up and lets a few goals in, it can spell disaster. The difference between a 90 percent and 85 percent save average adds up to a dramatic effect at the end of the season in terms of games won, and it can mean being relegated to the minor leagues. To stay in the NHL, goalies must fine-tune themselves and devote their full efforts to their job.

If there is a quality shared by the top goaltenders, it is their quickness. The art of blocking shots is about putting the pad or the blocking glove (also called the blocker) where the puck is heading before it gets there. To do so on a blistering slap shot, you have to have a lightning-quick response time. We begin this chapter by analyzing this aspect of goaltending with the help of physics. Following this is a discussion on the goalie's anticipated reaction to a shot and the cross-sectional area (the area of the net the goalie blocks).

Reflexes

Just before Islanders goalie Chris Osgood begins to move his glove to grab a fast slap shot, there is a brief moment—so short that it is barely visible on the slow-motion replay—during which he does nothing. This split second is his reaction time, and its duration is determined by how good his reflexes are. Before moving in any direction at all, Osgood has to figure out where the puck is going and how fast it will get there. Only then can his brain process the information and determine whether he should use his leg or his arms to stop the bullet.

One property of human reflexes is that we have very little control over them. Because our reflexes have evolved over millions of years as involuntary survival mechanisms, we are simply born with them and their quickness is genetically inherited for the most part. In other words, the time our body takes to react to a stimulus is not something we can improve with practice. Demonstrating this is quite simple. Just about everyone has tried the knee-jerk reflex experiment, something included in a routine physical examination. Sit down, let your leg hang freely, and have someone give it a gentle knock on the tendon just below your kneecap. Automatically your leg responds by kicking, and how fast it responds is not something you can concentrate on and control. Doctors use this test to determine a patient's state of well-being and alertness. Experts tell us that in healthy people, the time elapsed between the tap and the jerk is on average 0.05 seconds, or 50 milliseconds (1 millisecond is one thousandth of a second), but the response time can be shorter for some people. So when top NHL goalies like Osgood face a fast shot, they first rely on their innate ability to react quickly. Of course, taking little time to react does not guarantee a brilliant save, but it's a start. On close shots, quick reflexes are always critical.

Reaction times vary naturally from one person to another, but they also depend on, among other things, the type of stimulus sent to the brain. Humans react the fastest to sound stimuli—it takes from 140 to 160 ms to respond to them, compared to 155 ms for touch stimuli and 180 to 200 ms for visual stimuli. The difference is partly explained by the fact that it takes only 8 to 10 ms for the neural signal to reach the brain from the ear but 20 to 40 ms from the eyes.

Reaction times improve from childhood into the late 20s, but they progressively slow down from that point on. Reflex degradation then accelerates after a person reaches 70 years of age. For all age groups, men tend to have faster reaction times than women, and the difference is not reduced by practice. Pressing a button in response to a light signal takes on average 220 ms for men and 260 ms for women. As we might expect, fatigue and distraction have been shown to increase reaction times. Oddly, researchers also found that our reaction times are faster during exhalation than inhalation. Fitter people tend to have better reaction times, as do smarter people, although there are large variations within groups of similar intelligence. Finally, researchers have even found that punishment in the form of electrical shocks in response to slow reactions helps reduce reaction times.

Measuring Your Reaction Time

There are many ways to measure how good a person's reflexes are. The best and most sophisticated techniques involve electronic devices that measure the time needed to accomplish simple tasks. A subject may be put in front of a screen and told to press a key as soon as she sees a light bulb turn on. Computerized systems then measure the time elapsed between the two events.

You can still get a pretty good estimate of your reaction time without complicated apparatus. All you need is a long ruler (30 cm or more), the help of another person, and good old gravity. Ask someone to hold the ruler from the top so it hangs vertically and place your thumb and index fingers on each side of the ruler's bottom. Have your assistant let go without warning and then try to close your thumb and index finger on the ruler as quickly as possible. Because free-falling objects drop at a fixed rate, namely 9.8 m/s^2, there is a direct relationship between the distance d the ruler has dropped and the time t elapsed before you caught it (your reaction time). The formula we need to do the conversion is the same seen earlier (equation 3.1) for the puck trajectory:

$$t = \sqrt{\frac{2d}{g}} \simeq 0.45\sqrt{d}, \tag{5.1}$$

where d is in meters. Listed below are a few reaction times corresponding to falling distances of the ruler:

d (cm)	t (s)
5	0.10
6	0.11
7	0.12
8	0.13
9	0.14
10	0.14
11	0.15
12	0.16
13	0.16
14	0.17
15	0.18
16	0.18
17	0.19
18	0.19
19	0.20
20	0.20
21	0.21
22	0.21
23	0.22
24	0.22
25	0.23
26	0.23
27	0.23
28	0.24
29	0.24
30	0.25

Just like any other technique, even this simple method is not foolproof. You shouldn't watch your friend's hand to anticipate the drop (that's cheating). Also, if you catch it right at the beginning of the drop, or within a couple of centimeters, this is most likely not because your reaction time is lightning-quick but the result of a "false start," a timely, lucky move. To get an accurate result, take the average over several trials and use the table above to calculate your reflex time. Most people catch the falling ruler between the 15 and 25 cm mark.

How fast should your reaction time be if you aspire to become a pro goaltender? NHL goalies commonly face 80-mile-per-hour slap shots, so from 40 feet away how much time do they have? According to $t = d/v$, we find that they only have about 0.3 seconds! They also need time to move their glove and legs. So if an NHL goalie takes half that time to complete that move, he is left with only 0.15 seconds to make a decision on a shot. That reaction time is equivalent to catching the ruler at the 11 cm mark.

The Limits of Human Reaction Times

I consider myself to have fair reflexes. I can sometimes surprise myself by reaching down and grabbing a falling object before it reaches the floor. On the other hand, when playing with my cat I find it impossible to pull my finger away before she grabs it with her sharp claws. But felines are renowned for their quickness, a trait they have acquired over generations of hunting fast prey like birds and mice. They also have the ability to flip in midair and orient themselves to land on their feet, all in a split second. This brings us to wonder whether there are specific limits to animal and human reflexes.

From a physics point of view, the absolute shortest reaction time possible would be the time it takes for a signal to travel from the receptors located, say, at the tip of our finger, to the brain, and then back to the muscles so they can move in response to the stimulus. The absolute lower limit is on the order of a nanosecond (one billionth of a second); this is the time light—the fastest thing around, traveling at a speed of 300,000 kilometers every second—takes to cover a distance of about one meter. Because we know all human reaction times are much slower than that, it's not likely that the speed of the signal is the limiting factor. Even when humans are focusing on a simple task, like waiting for the pop of a starter pistol before running, their reaction times are quite a bit longer than a nanosecond. At the 2000 Summer Olympics, electronic devices used at track events to measure reaction times of sprinters showed that they are typically in the 0.15- to 0.25-second range. That is, once the sound signal reaches their ear, it takes about a fifth of a second before their foot pushes on the starting block. Just like goaltenders, sprinters need good reflexes in a sport where every hundredth of a second counts.

To understand reaction times better, we need to take a closer look at how the nervous system works. It contains billions of specialized cells called *neurons,* through which electrical signals propagate. Depending on the type of neuron, the signal travels at a speed that varies from 3 to 90 meters per second. This is much slower than the speed of light. So in the best of cases, the minimal time for the neural signal to travel one meter would be about 10 milliseconds. This is determined in part by how long it takes for the electrochemical signal to cross one neuron and jump to the next one. The transmission mechanism is a very intricate process that is limited by how fast a chemical reaction within each neuron occurs.

During an activity as complex as making a save in hockey, the reaction time needed to initiate the move doesn't simply boil down to the propagation speed of the neural signal: there is also a decision process that must take place in the brain, and this takes some additional time. Psychologists identify two types of reflexes in higher animals. The first kinds are called *unconditioned* and are innate. These are responsible, for example, for pulling your hand away from a hot stove. The brain doesn't have to think before it tells the hand to get out of there— the local nerve cells take care of that. *Conditioned* reflexes, on the other hand, take place when the brain makes a complicated but swift decision based on past experiences. Ivan Pavlov, the great Russian scientist and Nobel laureate for medicine, elegantly demonstrated this phenomenon with his famous dog experiment. His canines would salivate uncontrollably at the sound of a buzzer, knowing that a snack was on its way. The animals exhibited such behavior only after repeated trials, indicating it was a learned reaction.

Similarly, humans unconsciously respond to certain stimuli based on their past experiences. In a way, when Chris Osgood quickly reaches out to grab a puck, he is behaving like Pavlov's dog. Osgood's lengthy training has conditioned him; he knows there is satisfaction in making a save. His decision on how to move is not the result of a lengthy and complicated brain process, it is already "imprinted" in his mind. How quickly he reacts to a shot is partly determined by how reinforced the decision-making process is, hence the importance of regular training. Through his career, Osgood has seen hundreds of thousands of pucks, so his reflexes have been well trained to do a good job.

What exactly is the reaction time of a well-trained goalie? In the 1960s the Sport College in Toronto, then headed by famous hockey coach Lloyd Percival, conducted tests on professional goaltenders. They found that 0.2 seconds elapsed before a goalie moves his arm and 0.4 seconds elapsed before his legs got going.[1] In today's highly competitive NHL, these numbers might be smaller, but not by much. The 0.2-second limit is already very good, as you might have found with the ruler test. Simply pinching a ruler this quickly is hard enough, let alone making a complex decision on the puck's trajectory.

The 0.2-second minimum reaction time has a profound significance for the goaltender. It means that within a certain shooting distance from the net, reflexes are not sufficient. The puck would zip by before the goalie had a chance to do anything. This is why goalies are helpless against a shot deflected near the crease (the marked circular area directly in front of the net). If the puck is coming from that close with no warning, it's just too late to stop it. From the slot (20 feet away), the fastest NHL slap shot, at 100 miles per hour, would enter the goal in just 0.14 seconds. But there is still hope, otherwise a guy like Los Angeles's Felix Potvin might as well cross his arms and do nothing. Potvin can instead try to anticipate where the player might shoot and start moving ahead of time. If he's lucky and reads the play well, he still has a good chance to block the shot.

Anticipating a Shot

There are signs that tell a goaltender a puck is about to come his way. Apart from the obvious winding-up motion before a slap shot or a slap-wrist shot, the shooter's eyes and facial expression also indicate

---

1. J. Hunt, *Bobby Hull* (Toronto: McGraw-Hill Ryerson, Ltd., 1966).

where he will shoot. At that precise moment, the goalie must go into high alert and be ready to stop the puck.

Simply knowing that a shot is coming doesn't tell a goalie where it will be aimed at. Professional goaltenders learn to guess the direction of a shot based on how the puck is hit. For example, a wrist shot is more likely to go high than is a slap shot. Depending on how the stick is swung and the angle at which it hits the puck, a pro goalie can figure out approximately where the puck will go. Hints may be very subtle, but they are useful nonetheless.

Goalies can do even more than anticipate the direction of a shot. They often take an active role in determining where the player will shoot. A trick I have used in the past with some success is to entice the shooter to aim at a particular spot by purposely leaving an opening in the net. Usually I leave the glove-side partly unprotected—an opportunity too good to be missed by a player hungry for an easy goal—and so I expect the shot to come that way. If the player falls for it, I have the upper hand and can react quickly. Even if it means gaining just a fraction of a second, knowing where the puck will go is truly an advantage.

Of course, the other side of the coin is that a player can also trick a goalie into moving the wrong way. Hockey commentators sometimes talk negatively about a goalie making the first move, especially on a breakaway. When this happens it usually means the player managed to entice the goalie to go one way while he went the other, easily scoring in an open net. During a one-on-one confrontation, shooters use body language, faking head and stick moves to trick the goaltender into moving out of position. For the goalie, making the first move can then be dangerous. If a goalie wrongfully anticipates a low shot and kneels down to cover the bottom of the net, he might end up watching the puck dash over his shoulder. This is the main problem with the popular butterfly technique, actually. Coaches instruct their players to shoot high on goalies like Colorado's Patrick Roy and New Jersey's Martin Brodeur because of their tendency to kneel down quickly, an integral part of the butterfly style. "Stand-up" goaltenders like New York Ranger Kirk MacLean may be tougher to beat on high shots.

The idea behind the butterfly technique, invented by goaltending legend Tony Esposito, is to promptly block the bottom of the net

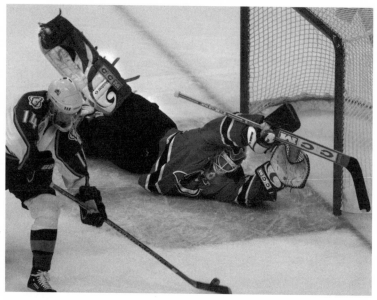

Figure 5.1.   New Jersey Devil Martin Brodeur prepares to make a desperate save on a backhander by Colorado Avalanche Dave Reid. CP Picture Archive (Kevin Frayer).

and still cover the upper areas with the quicker glove and blocker. As mentioned above, our legs tend to move and react slower than our hands, so the bottom of the net was typically the goaltender's weak point. Butterfly goalies are quick to kneel down and shut this entire area off, including the "five-hole," the opening between the goalie's legs. To encourage snipers to try to shoot into the five-hole, goalies also began wearing pads with triangular-shaped patches of white that create the illusion of a wider gap. These pads were introduced a few years ago and are now very popular.

Each style of goaltending has its strong and weak points, so it is a matter of finding what works best for a given athlete. There is no "one size fits all" kind of formula. Yet, the butterfly style has become incredibly popular over the last decade for a number of reasons. One of them is the great success of former Montreal Canadien Patrick Roy, who showed the world how effective it could be. Nevertheless, it took many years before people took notice; even Roy himself didn't often use the technique early in his career. Recently, when Roy surpassed

Terry Sawchuk's record 447 career wins, old footage of the mid-1980s Montreal Canadiens was shown during the tribute; it was surprising to see how much more Roy stood up back then. His style changed considerably over the years, until he had mastered his technique with remarkable efficiency. One of the most memorable moments was when Roy blocked 39 shots against the Boston Bruins to lead Montreal to a 5–2 victory in game four of the 1994 first-round playoffs, this after he'd convinced the team doctors to let him play while he was recovering from appendicitis!

Along with anticipating a shot, concentration is also very important in stopping pucks. A quick reaction time is useless unless it is backed up by a relentless focus on the puck. Just a blink of the eye or a moment of distraction, and the next thing you'll notice is the other team celebrating a goal. This is why NHL goaltenders try to focus on the puck itself rather than the puck carrier or other players. Sure, knowing where potential puck receivers are is helpful (peripheral vision is useful for that purpose), but the small black dot is what the goalie is interested in. Even at the amateur level, where the pace is much slower, I sometimes find it hard to peek around and locate dangerously positioned opponents in front of the net. By the time I look back at the puck handler, it might be too late.

What separates star goaltenders like Ed Belfour and Dominik Hasek from the rest of the crowd is their tireless focus on the game and their inner drive to win. It's their consistency and intensity, not so much their superior ability, that is key to their success. Coaches can rely on them at every game. But sustained concentration is not easy. It's hard enough to pay attention to a puck for 60 minutes, let alone a whole season or a whole career. A goalie can easily lose focus as a game gets out of hand or fatigue sets in. Even then, it's important to stay alert—there is still a possibility that the game can be won.

Most of the time, a goalie will anticipate a shot correctly and act swiftly to make a save. But what can one do when there's just not enough time to react, when players are buzzing around the crease? Frankly, all a goaltender is left with is luck. Well, luck and the hope that one's posture will not allow the puck to hit the net. In other words, the goalie will try to cover the net as much as possible to increase the chances of blocking the shot. This brings us to the all-important concept of cross-sectional area.

The Art of Looking Big

Felix "The Cat" Potvin of the Los Angeles Kings lies low with his legs spread apart when he waits for a shot (a posture referred to as "the stance") not to look cool but to do two things. First, by positioning himself this way he minimizes the time it takes for his legs and arms to reach out and cover all areas of the net. This is why goalies keep their glove high—it's ready to stop a shot at the corner. Potvin keeps his legs slightly apart to give himself a head start against pucks fired at bottom corners of the net. Second, he needs to maximize his blocking area, or, as physicists would say, his cross-sectional area. The larger he appears from the shooter's point of view (or, more precisely, from the puck's point of view), the greater his chances of blocking the shot. As we saw in Chapter 3, the uncertainty in shooting is quite high, even at the professional level. This means the goaltender can always hope the puck will hit him instead of making it through a small opening. By increasing his cross section, he reduces the player's chance of scoring. So putting his blocker or his glove in front of his body would be useless, because it wouldn't cover any additional area.

According to the NHL rulebook, with the exception of the skates and the stick, all goaltending equipment must serve the sole purpose of protecting the athlete. Clearly, competitive athletes have stretched the rules. Goaltending equipment seems to get bigger every year. Is this because shots are more dangerous than before, so the goalies need better protection? Just compare the size of the hand glove worn by Ken Dryden in the mid-1970s to the ones used today. In Dryden's days, it was barely larger than a baseball glove, but now gloves have much larger pockets along with wrist cuffs—these are there to deflect shots, but above all they are designed to occupy more space. Patrick Roy understood the concept of cross section when he attached strings between his arm and his waist a couple of years ago. This way, when he spread his arms apart the string would stretch and widen his jersey and dramatically increase his blocking area. It was fine for Roy, but when players began to grumble about goaltenders that looked more like flying squirrels, the league took notice.

The increasing size of goalie equipment resulted in a reduction in the number of goals scored. If this was good news for goaltenders, it was bad news for the fans, the NHL, and the game in general. The league eventually introduced rules to curb the "goalie equipment inflation," so to speak, and penalties and suspensions were enforced against violators. Pads are now limited to 12 inches in width, and the thigh protectors at the front of the pants must each be under 11 inches wide. Inserts that used to be inserted underneath the shoulder and clavicle protector to elevate the shoulder pads (like football players) were ruled illegal. Gloves were also limited to 19 inches in length, from the tip to the heel of the pocket. Needless to say, NHL goaltenders make full use of the permitted dimensions. And who knows, without such regulation, NHL goaltenders might have slowly evolved into giant puck-catching gloves that could open wide and cover the entire net!

Even within the gear limits allowed by the NHL, the fraction of the net that can be covered by the equipment alone is considerable. The net stands 4 feet by 6 feet, and the leg pads alone can block as much as 20 percent of the space. The glove and the blocker together take care of about 10 percent, and the chest protector and arms cover another 25 percent. Add all of this up, and 55 percent of the net is now sealed—and that doesn't take into account the stick, the mask, and the skates! Overall, a well-postured goaltender can block about two-thirds of the net when standing at the red line of the goal crease. From some angles, most of the net can be covered, as Fig. 5.2 shows.

There may be a limit on the size of the equipment, but there's no limit on the size of the goaltender himself. I sometimes wonder why no team has ever thought of using some kind of sumo-wrestler-sized goaltender to completely seal the net. There are men who are large enough—with the proper padding—to hermetically block the net, thereby offering an attractive 0.00 goal-against-average and a sure trip to the Stanley Cup final! But of course, a highly paid brick wall used simply to stuff the net would be a disgraceful spectacle, not likely to thrill the fans or the officials. Nonetheless, like the trend in players, bigger goalies seem to be in greater demand these days in the NHL. Goaltenders like Calgary's Roman Turek, at 6'3" and 225 pounds, would have been rare two decades ago, but not today. In general, it is not clear whether size is a definite advantage—since taller and heavier players do not usually have as much quickness and agility.

On a more serious note, if 60 percent of the net can be covered simply by dressing up as a goalie, couldn't anybody do a decent job

Figure 5.2.   From the puck's point of view, Patrick Roy blocks 80 percent of the net. Is it surprising, then, that his save percentage is around 90 percent? CP Picture Archive (Bill Janscha).

stopping pucks? After all, the 85 percent save average needed to stay in the league doesn't seem that far away from the 60 percent you supposedly start out with. A little moving around and the deal is done, right? The problem is, the first 60 percent is the easy part; covering the remainder of the net—and this is where players aim at— is much harder. Even if the other team misses the intended opening half the time, they'd still manage to score 15 goals in a typical game. You would need many excellent shooters on your team to even the score! Consistently blocking most of the shots requires positional play and speed, which is where athleticism and technique come into play.

Playing the Angles

Before taking his stance, the goalie first moves to the right place in front of the net. His chances are better if he stands at some distance ahead of the crease and centers himself directly in the path of the typical oncoming shot. Leaving the smallest possible window of opportunity for the shooter in this way is called playing (or covering) the angles. Figure 5.3 illustrates this tactic, which is aimed at minimizing the angle at which the puck can enter the net. By moving in and out along the line of direct shooting—a technique called *telescoping*—the goalie increases or decreases the angles.[2]

How far ahead of the net should the goaltender stand in order to completely shut the angles? That question can be answered with physics if we make a few approximations. First, we need to take into account how "wide" the goaltender stands and how far away the

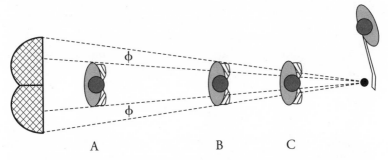

Figure 5.3. Telescoping is done by moving back and forth along the line of direct shooting. From the shooter's point of view, position A leaves a narrow margin of scoring opportunity, whereas B (the ideal position) leaves no chance at all. Though position C blocks all the angles, it leaves a wider opening behind.

2. For a detailed discussion on such goaltending techniques, see I. Young and C. Gudgeon, *Behind the Mask* (Victoria, British Columbia: Polestar Book Publishers, 1998).

shooter is. Of course, goalies are not rectangular in shape, so there might be more space at the top than at the bottom. For the purpose of this discussion, we will look at the lower angle, on either side of the goalie's pads. Suppose the shooter is a distance D away from the net and the goalie has a width w at the bottom. Then using simple geometry we find that the optimal distance d for the goalie to stand from the goal line is:

$$d = D(1 - w/6), \tag{5.2}$$

where 6 is the width of the net, so, consequently, all other variables are expressed in feet. Typically, w would be around 3 feet, so we obtain $d = 0.5D$. No matter how far away the shot is taken, then, the goaltender should be positioned at some fixed fraction—roughly half—of the shooter's distance from the net.

Equation 5.2 supposes the puck is coming from the center of the ice, namely along the line that joins the two nets. To the shooter, the goal area appears the largest along that line. Obviously, on shots from the sides, the net opening is narrower and the goaltender doesn't need to move that far away. For mathematically inclined readers, the proper formula in this case is $d = D(1 - w/[6\cos\theta])$, where θ is the angle of shooting relative to the center line (so that $\theta = 0$ when shooting from the front and $\theta = 90°$ from the far side of the net). This formula sometimes gives negative values for d, meaning the goalie can stand right at the net. As an example, a shot taken from the zone face-off dot means $\theta = 48°$, and an ideal telescoping distance would then be 7 feet, with $w = 3$ feet. (The telescoping distance would be 15 feet if the shot came from as far away but from the front of the crease.)

These formulas are only meaningful at close shooting range. If they applied all the time, the goalie would need to be at the blue line when a shot is taken from the center-ice! There's no need to venture any farther than about 15 feet from the net, because on a far shot there is more time to react. Playing the angles is not as critical at greater distances.

When a goalie doesn't move all the way up to the optimal position but is still well centered, from the puck's point of view there is a small opening angle ϕ on each side. This can be calculated using trigonometry:

$$\phi = \arctan\left(\frac{3(1-d/D)\cos\theta - w/2}{D-d}\right), \qquad (5.3)$$

where all distances are in feet. If ϕ is zero or negative, it means angles are completely shut.

This trick of zooming closer to and farther from the shooter has to be done with some caution. By moving away from the net, the goaltender leaves a wide unprotected area from the side. A shooter could then decide to pass the puck to a teammate for an easy goal. There is a compromise between minimizing the shooter's scoring chance and leaving a dangerous empty spot. But on a one-on-one breakaway (a goalie's nightmare), telescoping can be done without fear of being caught off-guard. At the moment the player breaks away, the goalie takes a few strides forward, then starts moving back as the opponent gets closer. As the loose player rushes toward the net, the goaltender also moves backward to keep the angles shut. In terms of skating speed, the goaltender has the advantage because, according to Equation 5.2, the goalie needs to be at a fraction of the distance away from the net and can therefore skate at about one-half the speed of the oncoming player. When done properly, telescoping covers the angles the entire time, and the player is left with two options: try to shoot through small openings like the five-hole, or go around. Going around is not a sure bet because the goalie will also move to the side to block. But the goaltender might also accidently make the first move—as mentioned earlier, players who decide to go around will often try to trick the goalie into sliding the wrong way. Here, the small fraction of goaltenders that catch with their right hand, like Montreal's Jose Theodore, are advantaged because breakaway players don't really have the time to think about which side is best and will often go to the wrong side of such players, namely the glove side.

One might wonder how a pro goalie moves around the crease with accuracy while keeping his back to the net. Doesn't he have to look back once in a while to see where he is? Usually there is no time for that, at least not in a fast-paced NHL game. Instead, a good goalie relies on experience and a sense of orientation. Goalies gauge their position by how far they have traveled. Knowing instinctively where the net is takes a lot of practice. Before a face-off, they touch the sides of the net with their stick and glove to make sure they are well centered. Ice markers—face-off circles and dots, the goal crease marking, and the logos—also help. This is why goalies who are used to playing on compact NHL rinks sometimes experience difficulties during international competitions on olympic-sized rinks. After a bad game in Vienna in 1996 against a young Slovakian team, Team Canada goalie Martin Brodeur had some explaining to do as reporters asked why he was out of position on so many shots. Some goaltenders, like Roberto Luongo of the Florida Panthers, are known for their impeccable positional play. But even they are not immune to miscalculations and can let the odd weak goal stray in once in a while.

Another aspect of covering the angles is changing position when the play moves around. A quick pass in front of the net, and you find yourself completely out of position! In such situations goalies must position themselves in a hurry while keeping a proper stance. To move quickly they need good skating skills and a solid sense of equilibrium. It is sometimes said that the goaltender should be the best skater on the team. There is some truth to this, for the goalie is the only one who needs to continually skate backward, forward, sideways, and spin around. Balance is a key element; losing it is a killer. As we saw in the chapter on skating, the goalie's skate has a flat blade that provides much more stability than regular skates. In spite of this, kneeling down and jumping back up repeatedly is difficult for anyone not accustomed to it. In physics, an "unstable position" refers to the state of a body ready to fall: its center of gravity is leaning outside its point of support. For the goaltender, support is provided by the skates and the stick. When a goalie loses balance, his or her center of gravity is not vertically in line with either, making it impossible to produce the acceleration needed to move swiftly in any direction.

Putting It All Together

Let's imagine the following situation: on a two-on-one fore-check, Owen Nolan of the San Jose Sharks crosses the blue line and decides to shoot at Red Wings' goalie Curtis Joseph. Having read the play well, Joseph moves 10 feet in front of his net, leaving a small unprotected angle of 1° (the equivalent of a 1-ft. opening on each side). Nolan fires the puck along the ice toward the inner side of the goal at a speed of 80 mph (or 115 ft./s). Will Joseph have enough time to make a leg save?

Let's analyze the play in slow motion. If we suppose that Joseph correctly anticipated the exact time of the shot and effectively reduced his leg reaction time from 0.4 s to 0.3 s, then the puck will have traveled $0.3 \times 115 = 35$ ft. before his leg starts moving. To travel the remaining 15 ft., the puck will take 0.13 seconds. Is that enough time for him to move his skate and deflect the puck? To find out, we need to know how fast his foot can move sideways. Since it is something no one is likely to have measured, we are left with guessing. Curious about this matter, I decided to determine my own lateral foot speed in our undergraduate physics laboratory using a computerized motion detector. After a few trials, I found that once my leg got going it took about 0.2 s to cover 40 cm. In other words, my foot moves sideways at a speed of 2 m/s. It would take me 0.15 seconds to move my skate by one foot and deflect the puck. That would be too late! But I'm not a highly trained professional athlete like Curtis Joseph—there is no doubt that he could cut that time by the 0.02 s he needs to get his foot to the right place early enough. So, the conclusion of our imaginary scenario is a frustrated San Jose Sharks captain returning to the bench, hoping he will be luckier the next time around.

We see in that example that Joseph would be expected to make the save because of his quick leg speed. But the anticipation of the shot is equally important, because on an 80-mph slap shot every 0.1 s wasted means the puck is 12 ft. closer to the net.

Chapter 6

THE GAME

Having analyzed various mechanical aspects of hockey, what can we say about the game as a whole? One may wonder at this point if we could apply the laws of physics to predict the outcome of a game or understand the way it is played. Unfortunately, we can't. If I could use a physical model to accurately tell me which team is going to win, I would have become rich betting on games! Because hockey players make complicated decisions, physicists will never be able to predict how the game will go. There are just too many elements involved. Even if hockey was played by robots obeying very simple commands, we'd still be faced with the problem of chaos: a tiny event, like a small air draft, may have dramatic effect later on in the game. In order for us to simulate a game accurately using physics, we would need a huge amount of information about each player and the environment—far too much for even the largest computers to process.

In spite of this, the business of hockey simulations is a growing, multimillion-dollar industry—thanks to computer games! These games are based on more or less accurate physical models for the players' skating motion, collisions, and puck trajectory. Developers of computer games were actually among the first to make extensive use of physical models to duplicate real-life athletes and vehicles. The most recent versions of virtual hockey games are sophisticated enough to let you trade players, customize them, arrange lines, choose a style of play, suffer injuries, have brawls, go through playoffs—and the list goes on. Graphics are also becoming more realistic each year, and the level of sophistication in the game is impressive. Each player has a certain level of scoring ability, endurance, speed, and shooting accuracy, just like in the real world. You can adjust the player's parameters and turn your roster into an all-star team while filling your opposing team with a bunch of wimps.

Virtual hockey leagues are popping up like mushrooms on the Internet. Participants take on the role of team owner and periodically pit their team against other teams. Owners trade players, arrange their teams and lineups, and draft new players every year. The simulation software, maintained on a server somewhere, is fed the configuration of each team before the game is "played." Results and analysis are then posted on a web site, just like news on real games appears in the papers and on TV. At the end of the season, virtual playoffs are organized and a champion is crowned.

The next logical question is, Can any of this software help predict the outcome of real NHL games? It is a valid question, but, once again, in order to be effective the program would need an awful lot of information about each team and the state of each player, all of which is obviously not available. Nonetheless, some limited information might be sufficient to give clues about who might win and by how much. Like weather forecasters, who rely on limited information about weather patterns, temperatures, and atmospheric pressures, we could eventually see the appearance of "sports forecasters" who would tell us the probability of a team winning based on sophisticated computer simulations. Who knows?

Another interesting mechanical question about hockey is how to rate the efficiency of different styles of play. This is of great importance to coaches, who must decide which players should play together and establish a different game plan from opponent to opponent. Creative coaches constantly experiment with new strategies. New Jersey head coach Jacques Lemaire's infamous "neutral-zone trap" in the mid-1990s—which served him so well in their 1995 Stanley Cup win—was a good example of an old defensive system that was revived, tweaked, and perfected until it became a powerful weapon. Even Scotty Bowman and his mighty Red Wings weren't able to escape the dreaded trap during the playoff final. The same trap system had been used with some success by the Montreal Canadiens in the 1970s, but it had not been exploited to the extent it was by Lemaire. The idea is to purposely leave a player in the middle zone to check opponents and prevent them from entering the offensive zone at top speed. This strategy was later blamed for lowering the number of goals scored and making the game dull to watch. Referees were told to more severely enforce the rules of clutching and grabbing in order to discourage the system's spreading throughout the league.

Hockey is a game that is constantly evolving. Coaches and players try new tactics, and some of them work better than others. Rules are sometimes changed simply to make the game more interesting. In the early days of hockey, a team consisted of a *goaler,* a *point,* a *cover point,* a *rover,* a *right wing,* a *center,* and a *left wing,* names that were borrowed from soccer. The point and cover point were the modern equivalent of the defensemen, except they lined up behind each other. This configuration lasted until someone realized it would be more effective for them to patrol the ice side-by-side. The rover was the middleman and the key player on the team; whoever played this position was usually the fastest skater. The rover was eventually dropped for economic reasons, but the position would have died sooner or later—seven men on each side certainly clutters the ice. Some fans today claim that four-on-four match-ups, like those that occur in overtime, are more fun to watch.

Pulling the goalie in order to add an extra attacker is also a recent invention—it was first tried about 50 years ago. During the dying minutes of the game, when trailing by one or two goals, a coach will sometimes replace his goalkeeper with an extra attacker to increase the chances of evening the mark. This typically happens near the one-minute mark of the third period when there's a single goal deficit or earlier when the difference is two goals. In a typical game, there will be an average of about 30 shots, or one shot every two minutes. By leaving the net empty one minute before the buzzer, the losing team is taking a moderate risk that no shot will be fired at their goal, although the opposite team will try harder to shoot from afar and put the last nail in the coffin. In 1950, one bold coach in Vancouver pulled his goalie 14 minutes before the end of the game, as his team trailed 6–2. The gamble paid off, and his team tied the game! But in today's NHL, no one would dare such a risky move.

Coaching and directing a team is done based on human experience; it is a world where physics and computer simulations are not helpful, at least not yet. There are physical reasons why, say, the configuration of three forwards and two defenders works better than four forwards and one player on defense. But the reasons behind this come from a large number of parameters, such as the size of the rink and the spatial range of control of each player, which in turn depends on their skating speed, shooting speed, size, weight, and so on. Trying to figure this all out scientifically would be a difficult task, albeit not impossible. If we did succeed in establishing a computer model for the best system of play, it certainly would be a great coaching tool. Lineups and game strategies could be altered according to the makeup of the opposing team and environmental parameters like rink size.

To hockey fans and analysts, the most important thing is the bottom line—who will win? Even today, all we can do is compile tons of statistics. But statistics go only so far, as hockey analysts know all too well. There's always an element of chance. Although hockey is not as statistics-friendly a game as baseball, huge amounts of hockey numbers are stored in databases every year. Some data are wrongly used by analysts, and others are probably irrelevant. There are many ways statistics may be used, all with various degrees of relevance.

This chapter uses the statistical method to address three questions. First, we will calculate the probability of a team winning. Second, we'll see what happens when a team plunges into a losing streak. Finally, we'll find out how good NHL hockey players really are compared to the rest of us—or, put another way, what is the likelihood of a person reaching the professional level in hockey? Above all, the aim of the exercise is to show two very different ways statistical analysis can be applied to understand hockey and sports phenomena in general. The methods used in this section can be found in any standard textbook on statistics.[1]

The Odds of Losing versus Winning

A good way to judge the strength of a team is by the number of games they have won. For example, the Dallas Stars of 1996–97 had a record

---

1. J. S. Milton and J. C. Arnold, *Probability and Statistics in the Engineering and Computing Sciences* (New York: McGraw-Hill, 1986).

of 48 wins, 26 losses, and 8 ties, for a total of 82 games played. In retrospect, we can say that the chance of the the Stars winning any given game that season was $48/82 = 58.5$ percent. The remaining 41.5 percent is the team's probability of not winning (that is, either tying or losing). In general, if a team has had w victories out of N games, the chance of winning any particular game is $p = w/N$, and the chance of not winning is $1 - w/N$.

Of course, average probabilities like these apply to randomly selected opponents. When Dallas meets a weak expansion team, we expect p to be greater than 58.5 percent. Game theory says that if two teams with p_1 and p_2 probabilities of winning meet, the odds for team 1 to win are $p = p_1/(p_1 + p_2)$ and $(1 - p)$ for team 1 not winning. This supposes all teams have played against each other equally, which is not quite the case in the NHL since teams are more active within their own division. Therefore, this formula is more accurate for teams in the same division, as they meet the same opponents an equal number of times.

Winless Streaks

Over the course of a regular season, every NHL hockey team, good or bad, goes through difficult periods. It is not uncommon for an organization to have a series of four, five, or more games in a row during which nothing seems to work. In these times of hardship, coaches and hockey analysts theorize about what is going wrong. Are things falling apart because of injuries, a lack of leadership, poor goaltending, or just plain bad luck? Sometimes the reasons are obvious—for instance, a star player may be sidelined because of injuries. At other times the answer is hard to find, in which case people often blame the demoralizing effect that a long winless streak has on the team's psyche. They say the team spirit is low, frustration is high, and this is why things are spiraling down. Psychology always seems to be held up as the main culprit for the lack of synergy and intensity in the play. It seems so obvious.

Is this accepted wisdom really true? Couldn't it be that winless streaks are just a matter of bad luck, an unavoidable product of randomness? To find out, I decided to do an investigation of my own.

It all started a few years ago when my favorite team, the Montreal Canadiens, was under fire for their dismal performance and for missing the playoffs twice in a row, something fans hadn't seen for a long time. Questions were asked from all directions by fans and the media. In Montreal, where hockey is a religion, there was a spirit of unrest over how the team was performing. It is a situation similar to the relationship New York baseball fans have with their Yankees: based on the team's historical record, they are expected to do well, or at least participate in the playoffs. So when the "bleu-blanc-rouge" loses two games in a row, journalists and analysts alike go to great lengths to find out what the problem is and see who should get fired for it. Meanwhile, if they are on a winning streak, the Canadiens are suddenly heralded as Stanley Cup contenders! This media scrutiny takes its toll on players, and, because of it, some prefer to play in cities like Phoenix or Denver. Although hockey is not as popular in these areas, at least players get a break from the media pressure. Annoyed by this quick-to-blame attitude, I decided to do some calculations to see if series of losses (or, more precisely, series of ties and losses) are statistical by nature. The idea is simple: just as flipping a coin will sometimes produce five tails in a row, any team can be unlucky enough to lose several games in a row—after all, there's a chance of losing every time they play.

To test my hypothesis, I compiled a few statistics from the 1996–97 NHL season. For a total of 12 teams, I recorded the number of winless series and classified them according to their duration. Then I computed the average duration of all streaks and compared it with what would be expected if winning and losing was just a matter of probability—like flipping a coin. How does one calculate the theoretical expected length of the average winless streak? It can be done based on the end-of-year record of a team. As mentioned above, if a team has had w victories out of N games, the chance of winning any given game is $p = w/N$, and the chance of not winning is $1 - w/N$.

From this probability of winning, the length of the average winless streak can be calculated mathematically. There are two ways to do this. First, you can grab a pen and paper, solve a bunch of equations, and come up with a nice formula. As it turns out, the calculation is cumbersome because of restrictions that are imposed (see Appendix 6 for details). For example, a winless streak longer than

82 games (the number of games in a season) must be ruled out. The other method is less analytical and much simpler. It involves a short computer program and the so-called Monte Carlo method of calculation. The technique is named after the town in Monaco famous for its casinos and gaming facilities. The Monte Carlo method was developed when mathematicians noticed that random events could duplicate real physical phenomena. The dartboard example is perhaps the most famous: the number of darts landing on a specific section of the board is indicative of the area of that section. This is because the chance of a dart falling inside a particular section is proportional to that section's area. If the area is circular, the dart trick can be used to approximate the value of π! In physics, the Monte Carlo method has been used to simulate complicated phenomena in quantum mechanics, thermodynamics, and nuclear physics. The first application occurred during the Second World War, when the method was used to understand and predict the random diffusion of neutrons inside the fissile material of a nuclear bomb.

How can the Monte Carlo method be applied to hockey? We can do so by generating virtual hockey games with a computer (like a computer model of flipping coins) and compiling the results into statistics. More precisely, to calculate the average length of a winless streak, we tell the computer that there is a probability p for the team to win any given game. Then the computer uses a random number generator to "flip a loaded coin," so to speak, so there is a chance p of the hypothetical coin falling on "win" and a chance $1 - p$ of it falling on "not win." After doing this 82 times in a row (representing games in the regular season), the computer finds the average of all winless series. But simulating one season is not enough. The result will vary from one trial to another. The computer must play about a hundred thousand of these "virtual hockey games" (yes, I know, it sounds tedious, but computers don't mind) before it spits out an averaged result that is very close to what would be obtained from the exact calculation. The greater the number of trials, the better the Monte Carlo approximation. I compared the two methods and obtained very similar results for 100,000 random games.

Now that we have a simple mathematical method, we can compare the theory with the compiled statistics for the 1996–97 NHL season. Fig. 6.1 shows the Monte Carlo simulation (represented as a

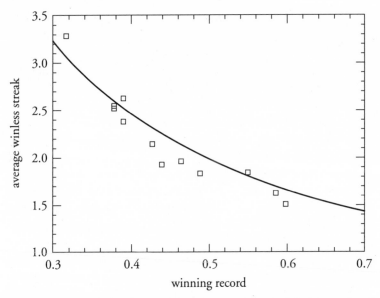

Figure 6.1. Expected duration of the average winless streak (solid curve) and actual duration (squares) for 12 teams in the 1996–97 NHL season as a function of winning probability. Most teams have shorter winless streaks than would be expected if winning and losing was a purely random process.

curve) along with the data points (squares) for twelve teams during that season. As expected, the average winless streak decreases with increasing *p*. In other words, the better the team, the shorter the losing streaks tend to be. Sure enough, all data points are very close to the theoretical curve, which indicates that winless streaks do follow a pattern of random probability. However, what is remarkable is that most points fall *below* the curve. This means that, for most teams, winless series are somewhat *shorter* than they would be if the process were purely random. Overall, data are about 5 percent below the curve generated by probabilistic theory, and this phenomenon is consistently observed and reproduced.

These results contradict the common assumption that a team spirals downward as a result of a long winless series. If a team played at its worst during losing streaks owing to psychological factors, then they would lose even more games in a row than normal and winless streaks would tend to be longer than expected. In reality, it seems to be working the other way around! Teams appear to play better after they have lost a few in a row. One could think of many reasons why this is so. For one thing, there is the added pressure from the fans and the coaches. Professional hockey players are motivated by more than just money. Delivering their best is a matter of pride and personal satisfaction. When hungry for a win, most of them will step up to the task. Therefore, teams that play a so-called demoralized opponent who has lost three in a row should beware!

Of course, my statistical compilation does not completely rule out the psychological element in hockey. There are times when teams are no doubt fired up, like when Mario Lemieux made an unexpected return on December 27, 2000, to his Pittsburgh team after nearly four years on hiatus. Although they had been playing poorly for quite a while, the Penguins shutout the Toronto Maple Leafs 5–0 to celebrate Lemieux's first game back. Trading a star player who has been underperforming or bringing in a new coach can also shake a team up. Other factors like injuries or being in a tight race for a playoff spot also affect performance. What our results show is that, by and large, most teams produce slightly better results after they have had a streak of bad luck. And there might be more to the story than that. Perhaps beyond a certain number of losses the situation is reversed. Weaker teams may experience unusually longer winless series compared to strong teams. To find out, more statistical analysis would be needed.

There is a great deal to be learned from what goes on when teams experience difficult times. Because it happens to every team at one point in their history, learning how to get out of the gutter is valuable knowledge. As strange as it might sound, in recent years I've become more fascinated by poor teams than by great ones (perhaps because I'm a Canadiens fan)! Instead of trying to understand what is behind great teams like the Flyers and Canadiens of the mid-1970s, I have become interested by the worst teams in NHL history, teams like the Ottawa Senators and San Jose Sharks of the 1992–93 season. While no one writes books on them, I think there's something to be learned from these examples.

That having been said, what do we know about winning streaks? At first, we might think that if losing streaks are shorter than normal, winning streaks should be longer. In fact, without doing any compilation, I can assure you that they are actually *shorter* than normal. The reason is not complicated. If we group lost and tied games throughout a season, then what remains—the wins—will also be grouped. Thus, increasing the length of losing streaks increases the length of winning streaks, and vice-versa. So if winning streaks in the NHL are indeed slightly shorter than what they would be if left to pure randomness, can we blame psychology—in the form of, say, overconfidence—for it? It's a possibility. The extra effort put in during difficult times and the careless play that creeps in when things are going well may work together to achieve series of both wins and losses that are shorter than expected. In some way, it's as if, when flipping a coin, there is a better chance of it falling on heads if it fell on tails the previous time. (Ironically, even though coin-flipping *doesn't* follow such a rule, many people think it works that way. The truth is that it has an equal chance of falling either way each time you flip it, no matter what the previous outcomes.)

I should point out that our analysis applies to predictions over the long run. Therefore, results are to be compared with data accumulated over the whole year. Of course, we might expect a shorter or a longer winning streak depending on who the opponents are. When a strong team like the Detroit Red Wings goes on a road trip to meet a bunch of weak expansion teams, they have a better chance at a winning streak than if opponents were selected at random.

In conclusion, before NHL team administrators start firing members of their organization as soon as things turn sour, they should consider the fact that winning and losing is also a matter of luck. Removing (or hiring) people may bring temporary motivation to the team, but it does not take away the element of chance that is always there. Even the best teams experience prolonged periods of bad luck. After all, isn't this unpredictability why we follow hockey in the first place? There is always a chance that the underdog will upset the Stanley Cup champions, and this is why every game is worth watching.

Meanwhile, here's a tip for hockey fans that wager on the outcome of games: bet on the team that is in the middle of a losing streak (or against the team that seems to be on a roll). This is actually when the losing team has a better chance of winning!

Against All Odds

Who has never wished to become a sport star? For many youngsters growing up in Canada, playing hockey in the streets or on frozen ponds, making it to the NHL someday would be a dream come true. Fueling those ambitions is the praise of parents who think their kid has the talent to be the next Wayne Gretzky. But for the vast majority of these kids, the dream of playing in the NHL will remain just that, a dream. And of course, much of the same can be said of girls aspiring to Olympic hockey greatness. Among the millions of amateur hockey players around the world, only a selected few reach the very top. As a result, sooner or later, high ambitions vanish and reality sets in. We understand that there are many young players out there who are as good or better than us. Those who have above-average talent may for a while seriously entertain the idea of becoming a pro, but one day they too meet their match. Among those who stand out from the beginning and climb the ladder to the university level or major junior teams, most will hit the wall at some point. It takes quite a bit of skill, determination, and, to some extent, luck to be scouted by a big league team. And even then, chances are that the new recruit will be a B-grade player who ends up famous only in his hometown! Considering the odds, it would appear that the chance for any kid to become a true star of the category of Pavel Bure or Joe Sakic is as unlikely as winning a jackpot twice in a row.

Just how small are the odds, and how can we estimate them? In science, odds are calculated by dividing the number of successes by the total number of attempts. Likewise, the odds of becoming an NHL player are determined by the ratio between the number of NHL players and the number of men who have had a chance to play hockey. So all we need are these two numbers. For our purposes, it's easier to use Canadian statistics, for two reasons: first, there are a significant number of Canadian-born players in the NHL, and, second, almost every Canadian boy (and plenty of girls—but since we're talking about the NHL, we'll stick to the boys) has put on a pair of skates and shot a puck at one point in time. A quick survey of ten NHL teams tells us that Canadians are a slight majority, with the percentage of Canadian-born players hovering around 55 percent.

In all, there are about 400 Canucks playing in the league. That seems like quite a bit, but out of how many candidates were these players chosen in the first place? Basically, the talent pool would consist of all the Canadian men in the same age group—roughly between the ages of 20 and 35—who have or had a genuine interest in playing hockey. The latest census estimates by Statistics Canada reveals that the population of Canadian men in that age group is 3.3 million. This means 1 out of 8,300 Canadian men have made it to the NHL. Realistically, we shouldn't count them all. Economic, geographic, and sociological factors are such that not all kids have the same opportunity or develop the interest to climb the hockey ladder, even if they are talented enough. Some remote rural areas have no arenas or hockey leagues, so they are not likely to produce the next Mario Lemieux. Also, some kids with a lot of talent are attracted to other sports or careers, or they don't get the necessary mentoring. All of this reduces the number of potential NHL candidates. Supposing that three-quarters of kids who are skilled enough go through the system as they should, the odds are now 1 to 6,000. In other words, 0.02 percent of all serious amateur hockey players eventually make it to the big league. Not as bad as we thought! In fact, it's far better than the chances of winning a big lottery.

Recently, many Canadians have expressed concerns over their decreasing influence in hockey on the world stage. There used to be a time when almost all NHL players were Canadians. Now, many top scorers hail from Europe and Russia. This has caused Canadian hockey experts to ponder the failures of their own training system. But the reasons may lie in the increasing popularity of hockey in the United States and abroad, as well as in the greater freedom of Russian athletes to leave the country, rather than a deterioration of the Canadian system. The more non-Canadian candidates there are, the less chance Canadians have to make it to the top.

So how good are those 0.02 percent that make it to the NHL? Although we recognize that some sport figures are great, their greatness is not easily measurable, especially in a field like hockey where a range of subtle skills are involved. We sometimes tend to overrate the best players based on how much they earn or how many points they've scored. The problem is, society rewards sport achievements (and also achievements in other fields) in a very uneven fashion. Fans—and team owners—like to put their money only on the very best, so anything slightly less than that is not good enough. Just think of the Olympic athlete who finishes fourth, a fraction of a second behind the leader, and is never mentioned. This is the reason why earnings are not a good yardstick for true greatness. A star hockey player who earns \$8 million a year is not 16 times better in sheer ability than another who earns half a million. Rating players according to their point production is equally deceiving. If goals and assists scaled up linearly with skills, this would mean Wayne Gretzky was only one-third as good at the end of his career as he was at his peak, when he scored 216 points in a season. He might have lost some of his magic touch by the time he retired, but not that much. As in any other competitive sport, the number of goals an NHL athlete scores is not proportional to his absolute skills but rather depends on the *difference* between his and his opponent's skills. It also greatly depends on the ability of his teammates. In his Edmonton days, Gretzky was fortunate to have Jari Kuri, Mark Messier, and Paul Coffey on his team. Mario Lemieux was equally lucky to play with Jaromir Jagr at his side.

We will attempt to solve the puzzle of hockey greatness with the help of a few statistical concepts. In the eighteenth and nineteenth centuries, scientists noticed that many biological and physiological features on humans and animals followed a *normal* distribution. This distribution is the well-known bell curve many professors use to grade their students. On a normal distribution curve, most people are found near some average, and the further away we depart from this average, the fewer people we find. Mathematically, it takes the form

$$p(x) \sim \exp\left(-\frac{(x-\mu)^2}{2\sigma^2}\right), \qquad (6.1)$$

where $p(x)$ is the probability of scoring x, μ is the average (also called the *mean*), and σ is the *standard deviation,* which we'll explain later. As it turns out, natural talent and physical features are spread among us in a normal distribution curve. This holds true for height, weight, life span, intelligence, running speed, endurance, strength, and so on. A normal distribution is encountered in nature whenever the outcome depends on a large number of variables, in this case environmental factors and genetics.

Normal distributions are fully characterized by two parameters, namely the mean and the standard deviation, the latter being related to how wide the bell curve is. A property of the normal distribution is that 68 percent of all individuals falls within one standard deviation of the mean. These two parameters are all we need to know about the distribution. For the distribution of many (if not most) human physical traits and abilities, the standard deviations are close to 15 percent of the mean. For example, the standard deviation for adult height is about 12 percent of the mean. It is 18 percent for the distribution of adult weight, 17 percent for birth weight, 16 percent for IQ, and around 15 percent for maximum running speed. This means most people differ from person to person by about 30 percent for any given skill or physical attribute. In that regard, human beings are actually remarkably homogeneous.

In light of this, it would be a reasonable assumption to say that pure hockey skills—however we decide to measure them—also follow a normal distribution with a 15 percent standard deviation. If we devise a practical hockey test so that the average score is arbitrarily set at 100, we can assume that most players (68 percent of them, to be exact) will score between 85 and 115. Even without administering the test, we can estimate the number of players that will reach a certain score s or higher with Equation 6.1. We just need to compute the area under the distribution curve (or its integral) for all x greater than s to determine the fraction of people who will be found there.

According to this idea, a fair way to estimate a hockey player's ability is to determine how far from the mean he stands. For example, if someone is the best among 100 individuals (the odds for this being 1:100), we should expect that person to score around 135. In other words, one player out of every hundred should reach 135 or better. In the case of NHL players, who come around once every 6,000 candidates, the minimum score would be 155. For superstars like Wayne Gretzky and Mario Lemieux, who show up only every decade or so (they would be one out of 500 NHL players, or one out of 2.5 million amateur hopefuls), the minimum score would be between 170 and 175, meaning they are 75 percent better than the average player. (See Appendix 7 for details.)

Some readers will wonder how it can be that NHL players are only about 60 percent better than the average John Doe. While it's true that it sounds like a small difference, a 60 percent difference in skills has a huge impact on the scoreboard. If you skate 60 percent faster, are 60 percent more agile, 60 percent stronger, and have 60 percent better coordination and stamina, you too could join the pros! But before the average hockey player runs to the gym with this total in mind, he should know that improving his game in all aspects by more than half is virtually impossible. Even 20 percent would be quite a feat. We are all born with a fixed genetic baggage and a certain amount of ability, and there is just so much we can improve upon. Improvements are usually small, in relative proportions.

The suggestion that there exists only a small difference between the skill of all hockey players—professional and amateur—tends to agree with observation. For example, when I attend a university hockey game, I don't find a huge difference between the university players and those in the NHL. Sure, university players are not quite as fast and intense as the professionals, but—although tickets for an NHL game are 10 times more expensive—the level of play is probably only about 20 percent better. Differences are even harder to notice when comparing the NHL with semiprofessional leagues like the American League, where the NHL "farm teams" compete. When a semipro player is given a chance to play in the NHL but sent back after a fruitless couple of games, it doesn't mean he is much less skilled than his teammates. Nonetheless, the slight difference is crucial. The very best players are a precious commodity because their slight edge and the extra effort they provide makes the difference at the end of a season.

Appendixes

Appendix 1: Units and Constants

Unless stated otherwise, equations in this book use the International System of Units, namely the meter, kilogram, second, joule, and newton. The following table gives the conversions to other familiar units.

Length
1 in = 0.0254 m

1 ft. = 0.305 m

1 mile = 1,609 m

Area
$1 \text{ m}^2 = 10^4 \text{ cm}^2$

$1 \text{ in}^2 = 6.45 \text{ cm}^2$

$1 \text{ in}^2 = 6.45 \times 10^{-4} \text{ m}^2$

$1 \text{ ft.}^2 = 0.0929 \text{ m}^2$

Volume
$1 \text{ liter} = 10^{-3} \text{ m}^3$

1 US gallon = 3.79 liter

Speed
1 m/s = 3.6 km/h

1 mph = 1.61 km/h

1 mph = 0.447 m/s

1 ft./s = 0.305 m/s

Mass
1 lb. = 0.454 kg

1 oz. = 0.0283 kg

Force and Weight (at 45° Latitude)
1 lb. is equivalent to 4.45 N of weight

1 kg is equivalent to 9.81 N of weight

Energy and Heat

1 calorie = 4.19 joules

Power

1 horsepower (hp) = 746 watts

Pressure

$$1 \text{ Pa} = 1 \text{ N/m}^2$$
$$1 \text{ atmosphere (atm)} = 1.031 \times 10^5 \text{ Pa}$$
$$1 \text{ lb./in}^2 = 6,900 \text{ Pa}$$

Angles

$$1 \text{ rad} = 57.3°$$

Constants

Acceleration due to gravity (at the Earth's surface and 45° latitude): $9.81 \text{ m/s}^2 = 32.2 \text{ ft./s}^2$

Heat capacity of water: 4,200 J/kg/°C

Heat capacity of ice: 2,220 J/kg/°C

Density of water at 20°C: 1,000 kg/m^2

Density of ice: 920 kg/m^2

Thermal conductivity of ice: 2.1 J · m/s/°C

Latent heat of fusion of ice: 3.4×10^5 J/kg

Appendix 2: A Mechanics Refresher

Mechanics is the branch of physics dealing with bodies in motion and the forces acting on them. It may not be the most exciting branch of physics—it's not as gripping as relativity or nuclear physics—but an awful lot of things can be understood with it. Most of the physics of hockey is described with mechanics.

Although mechanical phenomena are wildly diverse, they can all be explained using Isaac Newton's three famous equations, learned in every freshman physics course. This appendix reviews these laws and discusses a few other key concepts in mechanics.

A. Newton's Laws

1. *Law of inertia:* A body tends to stay at rest or keep moving in a fixed direction at a constant speed if no forces are applied to it (or when all the forces cancel out). In other words, linear motion and rest are the natural motion of objects.

2. *Law of acceleration:* The acceleration of a body is given by the net force acting on it divided by its mass. Mathematically, it is $a = F/m$ (or $F = ma$). I should emphasize that F is the *net* force—the total of all forces acting on the mass. Forces should be added vectorially whenever they are in two-dimensional or three-dimensional space, as will be seen in the next appendix. This equation is paramount in physics: from it we know the acceleration, and with the acceleration we can compute the velocity and the position of an object at any given time. With it, astrophysicists have predicted the trajectory of planets, asteroids, and comets.

3. *Law of mutual interaction:* When two bodies interact, the force on each body is the same but opposite in direction. When you push on the wall and it doesn't move, the wall is "pushing back" with the same force.

These three principles summarize our knowledge of mechanics. All other principles, such as the law of conservation of mechanical energy, are derived from Newton's laws. In order to be consistent in using formulas and equations, we need to use the SI unit system. When dealing with speeds and accelerations, we'll talk in terms of m/s and m/s^2. At the end of a calculation, you can always convert the results into more familiar units using Appendix 1.

B. Debunking Misconceptions
about Mass, Weight, and Force

The concepts of force and mass are central to physics. A force is simply whatever makes things move and accelerate, whereas the mass tells us how much matter a body is made of or how much inertia it has. The standard unit of force in physics is the newton (N). One N is the net force required to accelerate 1 kg of mass at 1 m/s^2. Therefore, 1N is equivalent to 1 kg · m/s^2.

It is important to understand that mass and weight, although they are related, are not the same. Weight is the gravitational pull on a body and is measured in newtons, but the mass is a quantity of matter and is given in kilograms or pounds. Your weight may change depending on where and how far away you are from the surface of the Earth, but your mass remains constant (unless you diet, of course). The acceleration caused by gravity, the free-falling acceleration, is $g = 9.8$ m/s². So according to Newton's second law, the downward force acting on a mass is mg, meaning one kilogram weighs 9.8 N near the Earth's surface.

C. Work and Energy

When you reach down, grab a puck, and lift it to your waist, you are spending energy. When you run up the stairs, you also consume energy by pulling your body mass upward. In physics, whenever a force F is applied over a certain displacement d, we say an amount of work W has been produced, given by $W = Fd$. Work corresponds to the energy spent to do the job and is measured in units of joules (J). The puck has a mass of 0.17 kg, so its weight is $mg = 0.17 \times 9.8 = 1.7$ N, and you need to spend at least $Fd = 1.7 \times 1 = 1.7$ J to raise it one meter from the floor. The force and the displacement need to be parallel. If they are not, only the component of the force along the displacement will contribute to work.

D. Kinetic Energy

A moving body can displace things and do work—it has what we call *kinetic energy.* The energy K associated with the motion of a mass m is given by $K = \frac{1}{2}mv^2$, where v is the velocity. It can be released to produce work or be converted into other forms of energy. For instance, when a moving billiard ball collides with a stationary one, it completely stops and all its energy is transferred to the other ball. When a hockey player crashes onto the board, his kinetic energy is converted into noise, heat, vibrations, and deformations.

Appendix 3: Playing with Vectors

Unlike mass and temperature, which are measured on a one-dimensional scale, force, velocity, and acceleration can exist in two or three dimensions. They are quantities that have direction as well as magnitude. In physics, we call them *vectors*. Because vectors have special properties, we need to be careful when adding them.

It is often useful to split a vector into components along certain axes, or directions. When something moves horizontally, like a puck slides on the ice surface, the forces that are relevant to the motion are along the plane of the ice. So if you only apply a vertical force, the puck doesn't move. The components of a force (or any other vector) are found using basic trigonometry. Fig. A3.1 shows the idea. The force F is oriented at an angle θ relative to the x axis. The components of F along x and y, labeled F_x and F_y, are the projections of the force along those axes and are obtained from the definition of the cosine and sine functions.

Newton's all-important second law connects the acceleration of an object to its mass and the net force acting on it. The net force is the *sum* of all forces acting on the object, including gravity, friction, and so on. We need to know how to add them in order to find the acceleration. The procedure for adding vectors is simple: add all the same components together. Fig. A3.2 demonstrates how it is done in the case of a puck hit with a stick. The puck is under the influence of four forces: the pushing stick (F), the upward pushing force of the ice (N), the friction force along the ice (f), and its own weight ($W = mg$). The net force giving the acceleration along x and y are found by adding the components of all forces along these axes. But

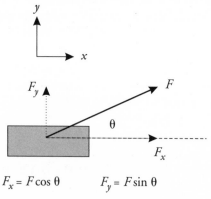

$$F_x = F \cos \theta \qquad F_y = F \sin \theta$$

Figure A3.1.   The vertical and horizontal forces on a puck.

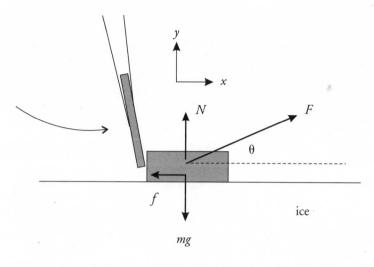

$$T_x = F \cos \theta - f \qquad\qquad a_x = T_x / m$$

total (net) force T

$$T_y = F \sin \theta + N - mg \qquad\qquad a_y = T_y / m$$

Figure A3.2. Forces on a puck as it is hit.

we need to be careful with the signs. The chosen direction of the *xy* system of axes determines the sign—whether each component is positive or negative. In the diagram, f and W are negative because they are oriented against x and y, respectively.

The same rules for finding components and adding forces hold true for other vectors, such as velocity, acceleration, and momentum. For example, in Chapter 4 we made use of the components of velocities when dealing with collision problems.

Appendix 4: Frictional Heating of Ice

The effect of friction is usually more important than pressure melting when it comes to skating, but it depends on a number of factors. One of them is the thermal conductivity of ice, labeled κ. The more conductive ice is, the more quickly heat dissipates when the skate blade rubs on it and the less heat per unit of volume it will bear.

The temperature rise ΔT is smaller as a result. We therefore expect to see an inverse relation between ΔT and κ, and Equation 1.4 confirms this.

Suppose the rubbing of a skate blade over some distance L makes the top layer of ice absorb an amount of thermal energy E. This increases the ice temperature by

$$\Delta T \approx \frac{E}{\rho VC}, \tag{A4.1}$$

with $V = Ad$, where A is the area of the blade contacting the ice and d is the dissipation depth of the heat into the ice.

The dissipation depth is estimated the following way. We suppose that over the time of rubbing, the heat absorbed by the ice flows through the area A and across a distance d. The heat absorbed causes a temperature gradient $\Delta T/d$ between the top layer and the ice just below. According to the definition of thermal conductivity, the energy flow is

$$\Delta E = \kappa tA\frac{\Delta T}{d}, \tag{A4.2}$$

where $t = L/v$ is the time it takes for the blade to completely sweep over one point on the ice. Assuming the heat keeps flowing and does not accumulate at one point, the energy flow given by Equation A4.2 is roughly equal to what is generated by friction (given by Equation A4.1). Putting all this together, we obtain

$$d \approx \sqrt{\frac{\kappa L}{\rho Cv}}. \tag{A4.3}$$

This formula says that: the faster the skater goes, the shorter d becomes and the greater the temperature rise, because the heat is dumped into a smaller volume of ice.

Appendix 5: Margin of Error during Shooting

Here we estimate the margin of error available to a shooter firing the puck from a distance d at a target (hole) of vertical dimension Δy

(see Fig. 3.8). The equations for a projectile moving in two dimensions are

$$y = v_{y0}t - \frac{1}{2}gt^2 \tag{A5.1}$$

and

$$x = v_{x0}t, \tag{A5.2}$$

where v_{y0} and v_{x0} are the initial vertical and horizontal velocities of the projectile, given by:

$$v_{y0} = v \sin \theta \tag{A5.3}$$

and

$$v_{x0} = v \cos \theta. \tag{A5.4}$$

Because the distance traveled is $d = v_{x0}t$, we can eliminate the time variable and write A5.1 the following way:

$$y = d \tan \theta - \frac{1}{2} g \frac{d^2}{v^2 \cos^2 \theta}. \tag{A5.5}$$

This relates the vertical position of the puck to the distance traveled, the shooting velocity, and the aiming angle θ. To estimate the margin of error on a small target, we use the definition of the derivative:

$$\frac{dy}{d\theta} \approx \frac{\Delta y}{\Delta \theta} \quad \text{or} \quad \Delta \theta \approx \left(\frac{dy}{d\theta} \right)^{-1} \Delta y. \tag{A5.6}$$

The derivative of A5.5 relative to θ is

$$\frac{dy}{d\theta} = d \left[1 + \tan^2 \theta - \frac{gd \tan \theta}{v^2 \cos^2 \theta} \right], \tag{A5.7}$$

yielding a margin of error for θ of

$$\Delta \theta \approx \frac{\Delta y}{d \left[1 + \tan^2 \theta - \dfrac{gd \tan \theta}{v^2 \cos^2 \theta} \right]}, \tag{A5.8}$$

where $\Delta\theta$ is given in radians (multiply by 57.3 to obtain degrees). But that's a nasty equation to use! Fortunately, θ is typically small ($<10°$), so $\tan^2\theta$ is negligible and $\cos^2\theta \approx 1$. For all practical purposes, $v^2 \gg gd\tan\theta$ if the puck is to reach the net before bouncing off the ice. Equation A5.8 is then reduced nicely to

$$\Delta\theta \approx \frac{\Delta y}{d}. \qquad (A5.9)$$

To find the margin of error on the shooting velocity, we use the same trick but derive A5.5 relative to v. The end result looks like:

$$\Delta v \approx \frac{\Delta y v^3 \cos^2\theta}{gd^2}. \qquad (A5.10)$$

Once again, because θ is usually small, it reduces to

$$\Delta v \approx \frac{\Delta y v^3}{gd^2}. \qquad (A5.11)$$

Appendix 6: Likelihood of Winning and Losing Streaks

This technical note is for statistics-savvy readers. We can develop an analytical expression for the probability of a hockey team of winning or losing several games in a row if we consider the process purely random. If w out of N games played have been won, the probability for winning any given game is $p = w/N$. A winning streak starts with a win and ends with a loss or a tie, the probability of which is $(1 - p)$. Therefore, the odds W_n of having an n-long winning streak are:

$$W_n = p^n (1 - p), \qquad (A6.1)$$

where n can take the values 0, 1, 2 . . . , with $n = 0$ meaning there's no win and just a single loss. If we repeat the process over and over, the average length L of a winning streak (including the game lost or tied to end it) is

$$L = \sum_{n=0}^{\infty}(n+1)W_n = \frac{1}{1-p}. \tag{A6.2}$$

Therefore, over the course of N games, the average number of streaks will be $N/L = N(1-p)$. Let X_n be the number of n-game-long winning streaks happening after a total of N games. We can find X_n by multiplying the total number of streaks by the probability W_n of having the n-streak:

$$X_n = N(1-p)W_n = Np^n(1-p)^2. \tag{A6.3}$$

However, we need to correct X_n because not all series have the same opportunity of happening. For example, during a regular NHL season ($N = 82$), a 5-game winning streak can only happen between the first and the seventy-eighth game, whereas a 2-game streak can happen up to the eighty-first game. The correction factor is approximately $(N - n + 1)/N$, giving

$$X_n = (N - n + 1)\, p^n\, (1 - p)^2. \tag{A6.4}$$

Finally, the average winning streak length S will be:

$$S = \frac{\sum_{n=1}^{\infty} nX_n}{\sum_{n=1}^{\infty} X_n} = \frac{2p - N(1-p)}{(Np - N + p)(1-p)}. \tag{A6.5}$$

This equation is accurate for $0.1 < p < 0.9$ when $N = 82$. Most NHL teams end the season with $0.25 < p < 0.75$.

The same formula is valid for the average length of winless streaks: we simply replace p with $p = (t + l)/N$, where l is the number of games lost and t the number of ties.

Appendix 7: Odds and Normal Distribution

Many biological features on humans occur according to the normal distribution, expressed mathematically as:

$$p(x) = \frac{1}{\sqrt{2\pi}\,\sigma}\, e^{\frac{(\mu-x)^2}{2\sigma^2}} \qquad \text{(A7.1)}$$

graphically, the normal distribution looks like Fig. A7.1. It is characterized by two parameters: the mean μ, around which the distribution is centered, and the standard deviation σ, related to the width of the bell curve. Equation A7.1 is *normalized,* meaning that the integral (the area under the curve) is equal to 1. The fraction of the population with a score of s or better is obtained by computing the area A under the curve from s up to infinity, or

$$A = \int_s^\infty p(x)\,dx. \qquad \text{(A7.2)}$$

This corresponds to the gray area in Fig. A7.1. Unfortunately, there is no analytical solution for this integral, so it has to be calculated numerically.

Since many human features and abilities are distributed over a standard deviation of the order of 15 percent of the average, we can suppose that natural hockey talent also follows the same pattern.

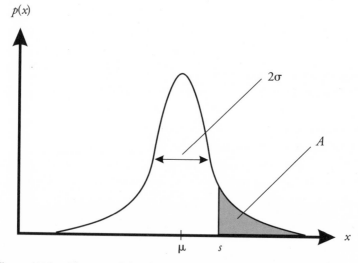

Figure A7.1. The normal distribution of mean μ and standard deviation σ.

If we arbitrarily set the average score of all hockey players at 100, the following table gives the odds of reaching various scores with a standard deviation of 15:

score	odds	
100	1:2	
105	1:3	
110	1:4	
115	1:6	
120	1:11	
125	1:21	
130	1:44	
135	1:100	
140	1:260	
145	1:740	
150	1:2,300	
155	1:8,100	(NHL level)
160	1:32,000	
165	1:140,000	
170	1:650,000	
175	$1:3.5 \times 10^6$	(NHL superstar level)
180	$1:2.1 \times 10^7$	

Glossary

Absolute zero: the lowest possible temperature, equivalent to $-273°C$ or $-459.6°$ F.

Acceleration: the rate of change in magnitude or direction of velocity.

Angular momentum: the momentum associated with the spinning of a rigid body.

Angular velocity: the spinning speed of a body, measured in units of the angle spanned (or number of turns accomplished) per unit of time.

Center of mass: the point at which all the weight (gravitational force) of a body appears to be applied.

Centripetal acceleration: the acceleration of a body caused by a continual change in the direction of its velocity. For an object on a circular path, the centripetal acceleration is directed toward the center of the circle.

Centripetal force: the force required to maintain a body in a circular trajectory.

Component: the projection of a vector (such as force or velocity) along a certain axis or direction.

Conservation of energy: a principle that states the total energy (including thermal, kinetic, potential, and radiative energies) of an isolated system stays constant.

Cross-sectional area: the area the silhouette of a body covers when viewed from a certain direction.

Deformation length: the maximum amount by which a body squeezes during a collision. In an elastic collision, this deformation is not permanent.

Density: the mass per unit volume of a substance.

Drag coefficient: a parameter proportional to the resistance encountered by a body when it moves through a fluid. It varies with the shape and texture of the body.

Elastic collision: a collision during which both the kinetic energy and the momentum are conserved.

Electron: a negatively charged particle that orbits the nucleus of an atom.

Force of impact: the average force of repulsion between two colliding bodies.

Friction: a force that opposes motion, such as rubbing or air drag.

Friction coefficient: the ratio between the friction force a body experiences on a horizontal surface and its weight.

Heat capacity: the energy needed to raise the temperature of a substance by one degree Celsius.

Heat (or thermal) conductivity: a quantity that measures how well heat can flow through a medium.

Inelastic collision: a collision during which the momentum but not the kinetic energy is conserved. A perfectly inelastic collision is one in which the two colliding bodies stick to each other (their final velocity is the same).

Kinetic energy: the energy associated with the motion of a mass.

Kinetic energy theorem: a principle that states the total energy expended by all the forces acting on a body is equal to its variation in kinetic energy.

Latent heat of fusion: the quantity of energy needed to melt one gram of matter without changing its temperature.

Moment of inertia: the inertia, or resistance, to rotational motion that a body possesses. It depends on the distribution of the mass about the center of rotation.

Momentum: the product of the mass times the velocity of a body; the total momentum of a system is conserved during collisions.

Normal distribution: a mathematical function represented by a bell-shaped curve. It can describe the spread of a certain trait or characteristic among a population of individuals or specimens. It is one of the most common types of distribution.

Normal force: the contact force between two bodies; it is oriented perpendicular to the surface.

Phase diagram: a pressure-temperature diagram giving the boundaries between the solid, liquid, and gaseous phases of a substance.

Potential energy: stored energy that can be released, like that in a loaded spring.

Power: the rate at which energy is spent or work is produced.

Pressure: the amount of force per unit area applied on a surface.

Probability: the ratio between the number of successes and the total number of attempts (or possibilities). It can take a value between 0 and 1 (or 0 and 100%).

Proton: a positively charged particle found in the nucleus of an atom.

Torque: the twisting force that tends to rotate a body. It varies with the force applied and the distance to the center of rotation.

Weight: the gravitational pull on a body, measured in newtons.

Young's modulus: a quantity proportional to the rigidity of a solid material (or inversely proportional to its elasticity).

Further Reading

D. Diamond, ed., *Total Hockey* (Kingston, N.Y.: Total Sports, 2000), is probably the most comprehensive reference book on the history of hockey and all NHL players and team statistics. Additional information about the science of hockey, including video clips and interviews, can be found on the Exploratorium web site at www.exploratorium. edu/hockey. *The National Hockey League Official Guide and Record Book,* published and updated every year by Total Sports, contains information similar to that in *Total Hockey,* although it is not as inclusive. Interesting, odd, and funny anecdotes about hockey, as well as the sport's history and the rules of the game, can be found in Brian McFarlane's *Everything You've Always Wanted to Know about Hockey* (New York: Pagurian Press, Ltd., 1973). *The Physics of Sports,* by A. Armenti (New York: Springer, 1992) is an assortment of papers on the physics of sports as varied as karate, tennis, and bowling. It also includes a few papers on the biomechanics of running and other track and field activities. *The Physics of Golf,* by T. P. Jorgenson (New York: Springer-Verlag, 1999) explains the science behind golf. A particularly interesting model for the mechanics of a golf swing is discussed. The technology of golf clubs and the aerodynamic forces on the ball are also examined.

Index